Infrared
Spectroscopy

Infrared Spectroscopy

James M. Thompson

PAN STANFORD PUBLISHING

Published by

Pan Stanford Publishing Pte. Ltd.
Penthouse Level, Suntec Tower 3
8 Temasek Boulevard
Singapore 038988

Email: editorial@panstanford.com
Web: www.panstanford.com

British Library Cataloguing-in-Publication Data
A catalogue record for this book is available from the British Library.

Infrared Spectroscopy

Copyright © 2018 Pan Stanford Publishing Pte. Ltd.

ISBN 978-981-4774-78-9 (Hardback)
ISBN 978-1-351-20603-7 (eBook)

To my daughter, Christine

Contents

Preface

This is an introductory text designed to acquaint undergraduate students with the basic theory and interpretative techniques of infrared spectroscopy. The author also believes that this material would be appropriate as an introductory text at the graduate level for those students lacking a background in the subject.

Much of the material in this text has been used over a period of several years in teaching portions of a one-semester course in materials characterization and chemical analysis. In addition to infrared spectroscopy, the course also includes a discussion of mass spectrometry and nuclear magnetic resonance spectroscopy. For the most part, the students in the course have had at least one year of organic chemistry. Thus, they have had at least a cursory exposure to the theory and interpretation of organic spectra.

In using this book in the course in materials characterization and chemical analysis, the author has devoted approximately 75 percent of the time to lecture discussions, with the remaining 25 percent devoted to the hands-on use of the various instruments at our disposal. However, depending on the needs and assessment of the instructor, the text could serve solely as a lecture text for a one-semester course in infrared spectroscopy or it could be used to teach the infrared portion of a broader course in materials characterization and chemical analysis.

In this undertaking, the author has tried to put together a book that is readable at the undergraduate or beginning graduate level. To do this, the text has been interspersed with many illustrations, examples, problems, and an extensive bibliography. In developing an introductory text, out of necessity, many more advanced spectroscopic concepts have been excluded. Despite these exclusions, the text should serve to give the student a solid background in the area discussed. It is hoped that those using the text would be sufficiently inspired to continue with more advanced study in the area of infrared spectroscopy. Students learn best by examples. Accordingly, the book includes many examples of the spectra representing a broad range of organic compounds. Thus, the text has been made more readable.

No author is entirely satisfied with his or her final works. Even as the final draft of the text is reviewed, there is an inclination to go back and expand, rephrase, and modify certain sections. Since this effort could continue for some time, what has been done is reluctantly submitted with the recognition that there are many areas for improvement. In a desire to improve upon this material, reliance is placed upon you, the students and the professors. Your recommendations and notification of the errors of submission and omission will be appreciated.

James M. Thompson

Acknowledgments

Several organizations and companies have been very generous in allowing the use of their infrared spectra in this text. This generosity has eased the writing of this text, and the author expresses his sincere appreciation to all involved.

Chapter 1

Some Fundamentals of Infrared Spectroscopy

1.1 Introduction

The infrared region occupies only a small portion of the electromagnetic radiation spectrum, lying between the more energetic ultraviolet region and the low energy radio frequency region (Fig. 1.1).

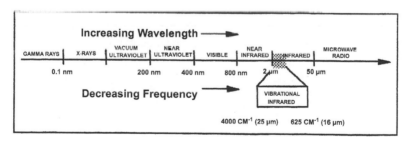

Figure 1.1   A schematic diagram of the electromagnetic radiation spectrum.

As we shall see, in the analysis of infrared spectra, it is common to report absorption bands in both frequency units and wavelength.

Infrared Spectroscopy
James M. Thompson
ISBN 978-981-4774-78-9 (Hardcover), 978-1-351-20603-7 (eBook)
www.panstanford.com

Reciprocal centimeters (cm^{-1}) or wavenumbers are units of frequency since they are equivalent to cycles per centimeters. Microns (μ) and micrometers (μm) are units of wavelength and their numerical values are identical. The recommended unit for expressing wavelength is reciprocal centimeters (cm^{-1}). In this text, the positions of infrared absorptions will be expressed in both reciprocal centimeters (cm^{-1}) and micrometers (μm).

On dividing 10,000 by the wavelength in microns, the wavenumber or frequency (in reciprocal centimeters) is obtained (Example 1.1).

Example 1.1

Convert 2.5 μm to wavenumbers (cm^{-1}).

$$1\,\mu m = 10^{-4}\,cm$$

$$2.5\,\mu m = 2.5\,\mu m\ \times\ \frac{10^{-4}\,cm}{1\,\mu m} = 2.5 \times 10^{-4}\,cm$$

$$\frac{1}{2.5\ \times 10^{-4}\,cm} = 4000\,cm^{-1}$$

Only that portion of the infrared region referred to as the vibrational infrared is of interest to most organic chemists (Fig. 1.1.) This region lies between 4000–625 cm^{-1} (2.5–16 μm). It is here that the frequency of radiation corresponds in energy to the natural vibrational frequencies of bonds in organic molecules. Of lesser concern to the organic chemist is the near infrared, which lies between 12,500–4000 cm^{-1} (0.8–2.5 μm) and the far infrared, 625–50 cm^{-1} (16–200 μm). *An infrared spectrometer measures the absorption of infrared radiation by the sample as a function of the frequency of the radiation.*

1.2 The Energy of Electromagnetic Radiation

As shown in the equation immediately below (also see Examples 1.2, 1.3, and 1.4), the energy of electromagnetic radiation is inversely proportional to wavelength (and therefore directly proportional to frequency). Thus, the frequency (in cm^{-1}) increases as the wavelength (in μm) decreases. Also, the larger the frequency, the more energetic the radiation.

$$E = \frac{hc}{\lambda} = h\nu; \text{ where } \nu = \frac{c}{\lambda}$$

λ = wavelength in centimeters

c = velocity of light (3.0×10^{10} cm/sec)

h = Planck's Constant (6.63×10^{-27} erg sec)

ν = frequency (sec^{-1})

Frequency may be expressed in units of reciprocal seconds (sec^{-1}), cycle per second (cps), or Hertz (Hz). Hertz has been recommended as the SI frequency unit; however, all three units are identical in numerical value and may be used interchangeable. As shown in the equation above, the frequency of electromagnetic radiation may be obtained by dividing the velocity of light by the wavelength (also see Example 1.2).

Example 1.2

Convert 15 μm to frequency in Hertz.

$$\nu = \frac{c}{\lambda},$$

where ν is the frequency (sec^{-1} or Hertz).
Since 1 μm = 10^{-4} cm, 15 μm = 1.5×10^{-3} cm.

$$\nu = \frac{3.0 \times 10^{10} \text{ cm/sec}}{1.5 \times 10^{-3} \text{ cm}} = 2.0 \times 10^{13} \text{ sec}^{-1} \text{ (or Hertz)}$$

Example 1.3

The frequency of a certain bond vibration is found to be 8.1×10^{13} Hertz. Convert this value to wavenumbers in cm^{-1}.

$$8.1 \times 10^{13} \text{ Hertz} = 8.1 \times 10^{13} \text{ sec}^{-1}$$

$$\lambda = \frac{c}{\nu} = \frac{3.0 \times 10^{10} \text{ cm/sec}}{8.1 \times 10^{13} \text{ sec}^{-1}}$$

$$\lambda = 3.7 \times 10^{-4} \text{ cm}$$

The wavenumber is the reciprocal of the wavelength (expressed in centimeters); consequently, it is only necessary to obtain the reciprocal of 3.7×10^{-4} cm to convert to wavenumbers.

$$\frac{1}{3.7 \times 10^{-4}} \text{ cm}^{-1} = 2702.7 \text{ cm}^{-1}$$

Example 1.4

Calculate the energy in ergs which corresponds to a photon of infrared radiation having a wavelength of 15 μm.

$$15 \text{ μm} \times \frac{1.0 \text{ cm}}{10^4 \text{ μm}} = 1.5 \times 10^{-3} \text{ cm}$$

$$E = \frac{hc}{\lambda}$$

$$= \frac{(6.63 \times 10^{-27} \text{ erg sec})(3.0 \times 10^{10} \text{ cm/sec})}{1.5 \times 10^{-3} \text{ cm}}$$

$$= 13.26 \times 10^{-14} \text{ erg}$$

1.3 Information That May Be Obtained from the Analysis of Infrared Spectra

The infrared spectra of organic compounds usually consist of an assortment of absorption bands (Fig. 1.2), all with different intensities and shapes and all occurring at wavelengths usually between 4000–625 cm^{-1} (2.5–16 μm). As we shall see, both the intensities and positions of the absorption bands are important in deducing the structural features of organic compounds.

Not all infrared absorption bands can be readily correlated, but those that can, often furnish answers to important structural questions such as:

(1) Is the compound a ketone, acid, amine, ester, etc.?
(2) Is the compound aromatic or aliphatic?
(3) If aromatic, what is the substitution pattern?
(4) Does the compound contain a double or triple bond?
(5) Is the double bond conjugated, isolated, or terminal?

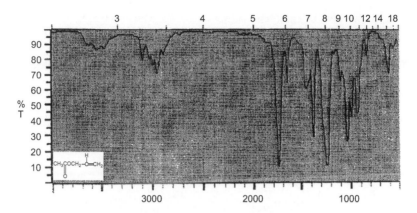

Figure 1.2 A typical infrared spectrum, showing the various absorption bands.

(6) Does the compound contain a tertiary butyl or isopropyl group?

(7) Which isomer best corresponds to the compound, *cis* or *trans*?

(8) To what extent is hydrogen bonding present, and is it intramolecular or intermolecular?

(9) Does the spectrum suggest a cyclic or open chain compound?

(10) Is the carbonyl group conjugated?

(11) If the compound is an alcohol, is it tertiary, secondary, or primary?

(12) If the compound is an ether, is it an alkyl–alkyl, alkyl–aryl or aryl–aryl ether?

(13) Are there few or many aliphatic methylene groups present in the molecule?

(14) Does the structure contain divalent sulfur?

(15) Is the carbonyl group chelated?

(16) Is the hydrocarbon cyclic or non-cyclic?

1.4 Comparison Techniques

Almost every organic compound has its own unique infrared spectrum, which may be considered as the "fingerprint" of the molecule. This characteristic allows for the identification of organic compounds by spectral comparison, i.e. if the infrared spectrum of an unknown compound completely matches one of a know compound, this is usually taken as conclusive evidence that the two compounds are identical. However, for spectral matching to be valid, the two spectra (know and unknown) must represent compounds of similar purity, taken in the same medium and under identical conditions. Otherwise a complete match-up will not result, even if both samples are identical. As an example consider the spectrum of triptycene, taken as a solution in chloroform and as a KBr pellet (Figs. 1.3 and 1.4).

Accordingly, if a spectrum is suspected as corresponding to a known compound, a comparison is best achieved by obtaining, if possible, an authentic sample of the suspected compound and running its spectrum in the same medium and under identical conditions as the unknown. If the sample is not available, one may proceed to search the chemical literature for its spectrum (see the references in the appendix).

In the absence of a matching spectrum, rarely is it possible to determine the exact structure of an unknown compound solely from an analysis of its infrared spectrum. Nevertheless, important structural information may be obtained in this fashion. When the

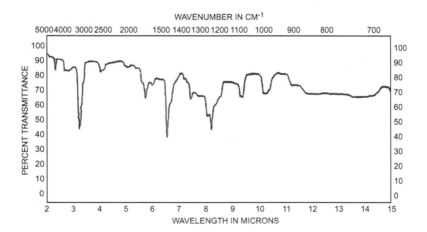

Figure 1.3 Triptycene as a solution in chloroform.

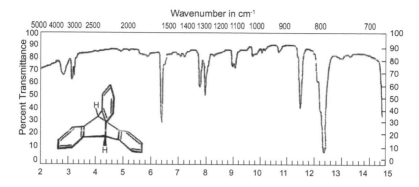

Figure 1.4 Triptycene (KBr pellets).

infrared information is combined with other spectral data such as ^{13}C and ^{1}H nuclear magnetic resonance (NMR), ultraviolet–visible (UV-visible), mass spectrometry (MS), and chemical data, the chances of a conclusive identification is enhanced considerably. The identity of an unknown compound may also be established by computer searching its infrared spectrum against libraries of digitized spectra.

1.5 Fundamental Vibrations

Chemical bonds are not rigid and inflexible entities, but are capable of undergoing a series of different vibrations. In this sense, molecules appear to behave as it they were balls connected to springs, with the balls representing atoms and the springs representing chemical bonds. Bond vibrations may be of several general types, consisting of asymmetrical and symmetrical stretching vibrations as well as in-plane and out-of-plane bending modes; all are referred to as fundamental vibrations.

Stretching vibrations involve motions by atoms in the direction of the bond, whereas bending vibrations (or deformations) involve motions resulting in corresponding changes in the bond angle. A brief description of some of the major fundamental vibrations are outlined below and illustrated in Fig. 1.5.

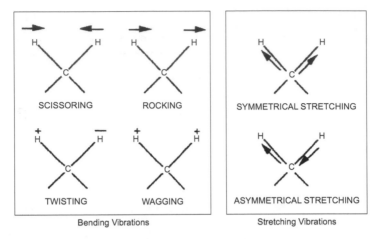

Figure 1.5 Illustrations of fundamental vibrational modes (+ and – denote movement above and behind the plane of the paper).

1.5.1 Stretching Vibrations

Symmetrical Stretching Vibration – involve similar bonds attached to a common atom, in which the bonds move away and toward the common atom in unison.

Asymmetrical Stretching Vibration – similar to symmetrical stretching vibration, but one or more of the bonds move away from the common atom while the other moves toward it.

1.5.2 Bending Vibrations

Scissoring: The bonds move back and forth with deformation of the valence angle, but the atoms remain in the same plane. The motion is similar to swinging both outstretched arms straight up toward the head and back, with the body as the common atom and the arms representing bonds. This vibrational mode is also called in-plane bending.

Rocking: The bonds swing right and left in unison, while the atoms remain in the same plane.

Twisting: The bonds twist out-of-plane around a common atom.

Wagging: The bonds swing back and forth in unison and out of the plane of the bond.

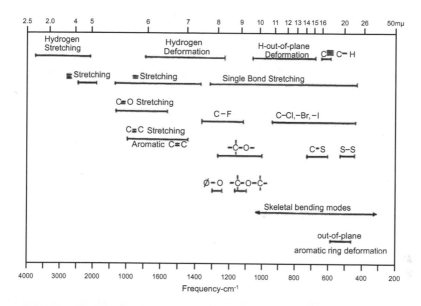

Figure 1.6 The regions of the infrared spectrum where fundamental vibrations normally occur.

Skeletal Vibrations: These involve motion of the entire molecule. Absorption due to these vibrations occurs between 1420 and 666 cm^{-1} (7–15 μm).

Generally, it takes more energy to stretch a bond than to bend one, thus stretching vibrations normally occur at higher frequencies and appear mostly in the left side of the infrared spectrum. The general regions for stretching and bending vibrations are shown below and in Fig. 1.6.

The General Regions for Stretching and Bending Vibrations

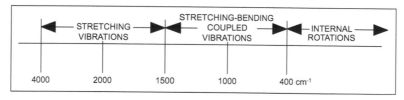

Every chemical bond has associated with it a characteristic vibrational frequency, the value of which depends upon the nature of the bond and its chemical environment. As a result, each bond will absorb infrared radiation of a characteristic wavelength. *The correlation of bond absorptions with their corresponding wavenumbers or wavelengths is the basis for the determination of structure features of organic compounds by infrared analysis.* If one or more identical atoms in a molecule are bonded to other identical atoms, the bonds will show both asymmetrical and symmetrical stretching vibrations. For instance:

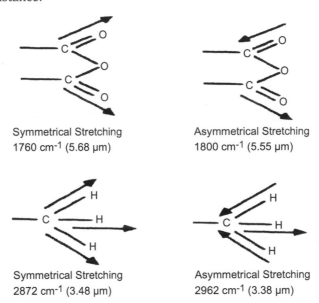

Symmetrical Stretching
1760 cm^{-1} (5.68 μm)

Asymmetrical Stretching
1800 cm^{-1} (5.55 μm)

Symmetrical Stretching
2872 cm^{-1} (3.48 μm)

Asymmetrical Stretching
2962 cm^{-1} (3.38 μm)

1.6 Non-Fundamental Vibrations

In addition to stretching and bending modes, the spectra of organic compounds often show certain non-fundamental vibrations which are the result of excitations to energy levels higher than fundamental vibrations. These bands are generally weaker than those of fundamental vibrations and are usually located in the low wavelength (high wavenumber) region of the spectrum. Non-fundamental vibrations may be classified as overtone, combination or difference bands. Overtone bands are usually located at positions which are 1/2, 1/3, or 1/4 the wavelength or a corresponding fundamental vibration. Occasionally these bands convey important structural information. One example involves the use of overtone bands in predicting the substitution patterns of benzene derivatives (Section 2.19). However, in most cases, the non-fundamental vibrations merely complicate the interpretation of the spectrum. Combination bands result from the coupling of two fundamental vibrations ($v = v_1 + v_2$); consequently, the position of these bands reflects the addition of two corresponding fundamental frequencies.

In some cases, a combination, overtone, or difference band may absorb near a fundamental vibration. When this occurs, there may be a decrease in the intensity of the fundamental vibration and an increase in the corresponding intensity of the non-fundamental vibration. This phenomenon is termed *Fermi* resonance and is sometimes observed in the spectra of carbonyl compounds. *Fermi* resonance may result from the coupling of two or more fundamental vibrations or from the coupling of fundamental vibrations with non-fundamental vibrations.

1.7 Predicting the Number of Fundamental Vibrations

Purely from a theoretical viewpoint, it may be shown that a linear molecule may have a maximum of $3n-5$ fundamental vibrations (where $n =$ the number of atoms in the molecule). For non-linear molecules, the expected vibrations decreases to $3n-6$. Oftentimes, not all of the predicted vibrations are observed in the infrared spectrum. Among the several factors that may contribute to their absence are the following:

- The absorption may be too weak to be observed.
- The absorption may coalesce with another absorption.
- The absorption may occur outside the region of the spectrum. For instance, the C–I stretching vibration appears between 602–500 cm^{-1} (16.0–20.0 μm) which is outside the range of some infrared instruments.
- Two or more absorptions may be degenerate (of the same energy), thus, occurring at the same frequency.
- There may be no net change (or very little change) in the dipole moment of the bond during vibration.

With respect to the dipole moment change, *it is required that bond vibrations be accompanied by changes in dipole moments if they are to result in the absorption of infrared radiation.* In fact, the greater the dipole moment change, the stronger is the intensity of the absorption band. Vibrations not accompanied by such changes are said to be infrared inactive. Infrared inactive vibrations are common to symmetrical molecules, including the diatomic molecules. For instance, based on the $3n–5$ rule, it is predicted that linear CO_2 should exhibit a total of four fundamental vibrations (two bending and two stretching vibrations); however the bending vibrations are of the same energy (said to be doubly degenerate) and one of the stretching vibrations is infrared inactive. Therefore, only two fundamental vibrations are observed in the infrared spectrum of the molecule (see Fig. 1.7).

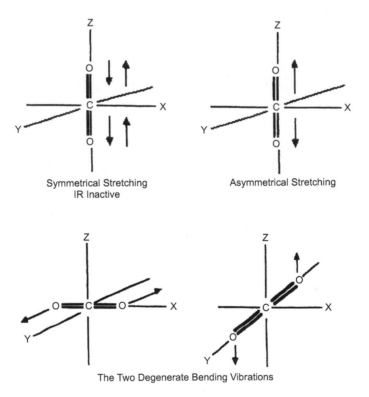

Symmetrical Stretching
IR Inactive

Asymmetrical Stretching

The Two Degenerate Bending Vibrations

Figure 1.7 The predicted fundamental vibrations of CO_2.

1.8 The Force Constant

Bond vibrations occur at quantized frequencies, i.e., only certain vibrational energy levels are allowed. The frequencies of these vibrations are determined, to a large part, by bond strengths, molecular geometry, and the masses of the connecting atoms. Bond strengths (and therefore vibrational frequencies) are influenced by factor such as (a) resonance, (b) hydrogen bonding, (c) electron withdrawing and donating effects, (d) hybridization, (e) steric interactions, and (f) the force constant of the bond. Therefore, depending upon the chemical environment and the other factors affecting bond strengths, bonds of the same general type may vary somewhat in the value of their force constants and their vibrational frequencies. Force constants may be viewed as a measure of the stiffness of the bond. Their values are directly proportional to bond strengths and may range between 3×10^5 to 18×10^5 dynes/cm, depending upon the nature of the bond (Tables 1.1 and 1.2).

As shown by the data in Tables 1.1 and 1.2, the following conclusions may be made about the relationship of force constant to bond strengths.

(1) Multiple bonds have larger force constants than single bonds.
(2) Force constants decrease in the order: $X \equiv Y > X=Y > X–Y$.
(3) The larger the force constant, the stronger the bond.

Using Hook's Equation (shown below) the stretching vibrations of common chemical bonds may be approximated from the force constant (Example 1.5). Conversely, the value of the force constant may be approximated from the bond's stretching vibration (Example 1.6).

Table 1.1 General ranges of force constants for some common bonds

Bond Type	Range (dynes/cm $\times 10^5$)
H–Y	3–7
X–Y	2–5
X=Y	9–12
X≡Y	15–18

Table 1.2 Force constants (and related constants) for some common bonds

Bond	Bond length (Å)	Force constant (dynes/cm × 10⁵)	Stretching (cm⁻¹)	Frequency (μm)
C–H	1.015	5.0	3038	3.30
N–H	1.012	6.5	3440	2.91
O–H	0.957	7.6	3648	2.71
S–H	1.335	4.0	2645	3.78
Si–H	1.48	2.7	2176	4.60
C–C	1.534	4.5	1128	8.86
C–N	1.47	4.9	1133	8.83
C–O	1.42	4.5	1051	9.62
C–F	1.39	5.6	1133	8.83
C–Cl	1.78	3.4	1246	8.03
C–Br	1.9	2.9	689	14.5
C–I	2.14	2.3	597	16.7
C=O	1.21	12.3	1746	5.73
C=C	1.337	9.8	1668	6.00
N=O	1.19	9.1	1433	6.98
S=O	1.43	10.0	1391	7.19
C≡C	1.20	15.6	2098	4.77
C≡N	1.128	17.5	2136	4.68

$$\nu = \frac{1}{2\pi c} \sqrt{\frac{K}{\dfrac{m_1 m_2}{(m_1 + m_2) 6.02 \times 10^{23}}}},$$

where ν is the frequency (cm⁻¹), c the velocity of light (3.0×10^{10} cm/sec), K the force constant (dynes/cm), $m_1 + m_2$ the sum of the atomic masses of the atoms connecting the bond, and $(m_1 m_2)/(m_1 + m_2)$ the reduced mass (μ).

By removing Avogadro's number from under the square root sign and evaluating the other constants, Hook's Equation above may be simplified to give

$$\nu\ (\text{cm}^{-1}) = 4.12 \sqrt{\frac{K}{\mu}}.$$

Example 1.5

Estimate the fundamental stretching vibration of an O–H bond which has a force constant, $K = 7.6 \times 10^5$ dynes/cm.

$$\nu \, (\text{cm}^{-1}) = 4.12 \sqrt{\frac{K}{\mu}}$$

$$= 4.12 \sqrt{\frac{7.6 \times 10^5}{\frac{16}{17}}}$$

$$= 4.12 \sqrt{8.01 \times 10^5} = 3702 \, \text{cm}^{-1} \text{ or } 2.70 \, \mu\text{m}$$

Example 1.6

The carbonyl stretching vibration of trans-2-hexenal appears at $1669 \, \text{cm}^{-1}$ ($5.99 \, \mu\text{m}$). From this information, estimate the force constant of the C=O bond.

$$\nu(\text{cm}^{-1}) = 4.12 \sqrt{\frac{K}{\mu}}$$

$$1669 \, \text{cm}^{-1} = 4.12 \sqrt{\frac{K}{\frac{12 \times 16}{12 + 16}}}$$

$$K = 11.26 \times 10^5 \text{ (in dynes/cm)}$$

1.9 Some Theoretical Concepts

As previously mentioned, each bond in a molecule undergoes a series of natural vibrations, all occurring at their own quantized frequencies which depend, to a large extent, upon the strength of the bond and its immediate chemical environment. Consequently, for most molecules, there exist a variety of bond vibrations with wavelength ranging 4000–625 cm^{-1} (2.5–16 μm). *When an organic compound is infrared radiated in this range, all bonds in the molecule whose vibrations are accompanied by dipole moment changes will absorb infrared at frequencies which match their own natural vibrational frequencies. These absorptions result in an increase in the amplitude of the vibrational motion which is detected and recorded by the infrared instrument. The result is a record of bond absorptions versus wavelengths or an infrared spectrum.*

Bonds absorptions are accompanied by a series of bond rotations. Because of this, the absorptions appear as broad to sharp bands rather than lines.

The stronger the bond, the greater is its vibrational frequency and the shorter the wavelength associated with its vibration. It follows that since single bonds are weaker than double or triple bonds they will generally vibrate at lower frequencies and their absorptions will appear at longer wavelengths. For instance, the C=O bond absorbs infrared radiation between 1870–1540 cm^{-1} (5.35–6.50 μm) due to stretching vibrations, while the weaker C–O bond absorbs between 1370–1000 cm^{-1} (7.3–10.0 μm). *Thus, it is possible to correlate the wavelengths of infrared absorptions with the stretching and bending vibrations of functional groups in organic compounds. It is this relationship that enables the analyst to determine structural features of organic compounds from the analysis of infrared spectra.* Example 1.7 will clarify the basic approach toward infrared analysis in which the frequencies of bond absorptions are correlated with molecular features.

Example 1.7

Stretching frequencies of O–H bonds in carboxylic acids and alcohols, (depending upon the extent of hydrogen bonding), appear as broad and strong absorptions between 3571–3195 cm^{-1} (2.80–3.13 μm). The C=O stretching frequencies of acids, esters, ketones, etc., appear as strong and conspicuous absorptions between 1870–1540 cm^{-1} (5.3–6.50 μm). Using this information, label Figs. 1.8, 1.9, and 1.10 as corresponding to an organic acid, ketone, or alcohol.

Answers

Figure 1.8 shows the broad and strong stretching absorption of the O–H bond near 3333 cm^{-1} (3.0 μm). The compound represented by this spectrum does not contain a carbonyl group since there is no strong absorption between 1870–1540 cm^{-1} (5.3–6.50 μm). Apparently, the spectrum corresponds to the alcohol.

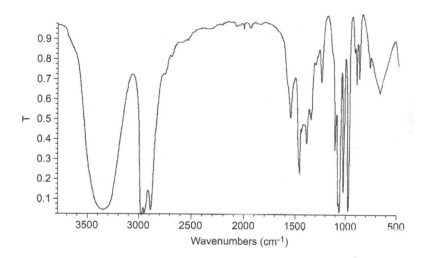

Figure 1.8

In Fig. 1.9, there is no broad and strong absorption in the 3571–3195 cm^{-1} (2.80–3.13 μm) region, thereby precluding the presence of an O–H group. It should be noted that the strong absorption near 2940 cm^{-1} (3.4 μm) corresponds to the aliphatic C–H stretching vibrations. Sometimes this absorption will overlap with the broad O–H stretching vibration; however, in Fig. 1.8, this is not the case. The strong absorption centered near 1712 cm^{-1} (5.84 μm) suggests the presence of a carbonyl group. Apparently, Fig. 1.9 is the spectrum of the ketone.

In Fig. 1.10, there is a strong O–H stretching vibration near 3125 cm^{-1} (3.2 μm) as well as a strong absorption near 1724 cm^{-1} (5.80 μm) which corresponds to the carbonyl group. This spectrum obviously represents the acid. In Fig. 1.10, it should be

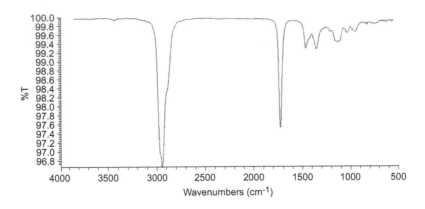

Figure 1.9

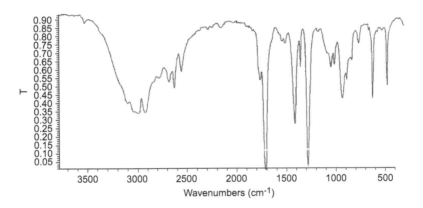

Figure 1.10

noted that the O–H stretching overlaps with the C–H stretching, but both absorptions are observed.

The position and appearance of the O–H stretching vibration in acids and alcohols depend upon the degree of hydrogen bonding of the O–H group. The greater the degree of hydrogen bonding, the broader the O–H stretching vibration between 3571–3195 cm^{-1} (2.80–3.13 μm). As you will see, in the assignment of bond frequencies, use is made of correlation charts and tables such as those shown in the appendix and by Fig. 1.21 and Table 1.4.

The compounds represented by the above spectra are
Figure 1.8: 1-propanol
Figure 1.9: 3-octanone
Figure 1.10: acetic acid

1.10 Basic Sample Preparation

1.10.1 Gaseous Samples

Most infrared spectra are taken on solid and liquid samples, but occasionally it is necessary to obtain spectra of gaseous samples. For this purpose, special cells, such as those illustrated in Fig. 1.11, are commercially available.

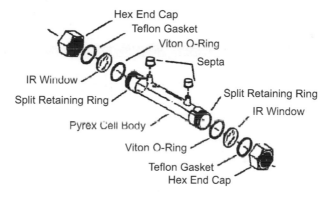

Figure 1.11 A standard infrared gas cell. *Courtesy of Spectra-Tech, Inc., a division of Nicolet Instrument Corporation, Stamford, CT.*

1.10.2 Non-Volatile Liquids

The spectrum of non-volatile or slightly volatile liquids may be taken neat (without dilution) by placing one or two drops of the sample between two identical sodium chloride plates. The salt plates are usually circular with a diameter of about 2.5 centimeters and a thickness of about 0.5 centimeters. In order to obtain a good spectrum, the sample compound must form a thin film between the plates. Because sodium chloride does not absorb in the vibrational infrared region, the entire spectrum is that of the sample. Based on the type of instrument, a spectrum may be obtained within minutes. Most modern infrared instruments require a background spectrum of the empty salt plates, followed by a spectrum of the sample between the plates. The background spectrum is automatically subtracted, resulting in the spectrum of the sample.

Salt plates should always be handled by their edges in order to avoid surface contamination. The plates should never be used to obtain spectra of aqueous samples or samples that will dissolve sodium chloride. Salt soluble materials will result in damage to the surface of the plates, thereby decreasing the transmission of infrared light. Furthermore, the spectra of samples contaminated with water will show the O–H stretching and bending vibrations of water. These bands, which are located near 3704 and 1627 cm^{-1} respectively (2.7 and 6.15 µm), could obscure the absorptions of the sample in these regions and may result in an erroneous structural assignments. For moist or aqueous samples, silver chloride, silver bromide, zinc sulfide or barium fluoride plates should be used. However, silver chloride and silver bromide plates may be damaged by amines and ammoniated samples. Properties of some common infrared materials are shown below in Table 1.3.

Table 1.3 Properties of some common infrared materials

Material	Infrared range (cm^{-1})	Properties
NaCl	40,000–625	Hygroscopic, water soluble, inexpensive, most commonly used material
KCl	40,000–500	Hygroscopic, water soluble
KBr	40,000–400	Hygroscopic, water soluble
CsBr	40,000–250	Hygroscopic, water soluble
BaF$_2$	67,000–870	Insoluble in water, soluble in acids and NH$_4$Cl, fragile
AgCl	10,000–400	Insoluble in water, corrosive to metals. Darkens on exposure to visible light. Should be stored in the dark
AgBr	22,000–333	Insoluble in water, corrosive to metals. Darkens on exposure to visible light. Should be stored in the dark
ZnS	50,000–760	Insoluble in water, common acids and bases, fragile

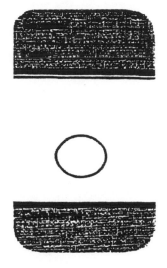

Figure 1.12   Sample card for obtaining the spectra of liquid samples.

The spectra of non-volatile and slightly volatile liquids may also be obtained by placing the sample on a thin layer of polymeric film such as polyethylene. The spectrum of the polymer is obtained first and then subtracted from the spectrum of polyethylene and the sample. Sample cards, such as the one shown in Fig. 1.12, may be used for this purpose.

1.10.3 Volatile Liquids

The spectra of very volatile liquids may also be obtained using salt plates, but the sample must be sealed between the plates to avoid evaporation of the sample. The simple but effective cell assembly shown in Fig. 1.13 is may be used to obtain spectra of volatile liquids.

As shown in Fig. 1.13, the cell assembly consists of two salt plates, a spacer, and a screw type holder to secure the cell. The spacers may be obtained in thickness of 0.025, 0.05, and 0.1 mm, allowing for variation in light transmission. In obtaining the spectrum of volatile liquids samples, one of the salt plates is placed on a flat surface and the proper spacers placed on the window of the plates. The area within the spacer is filled with two or three drops of the sample material and then sandwich between the two plates. The cell is

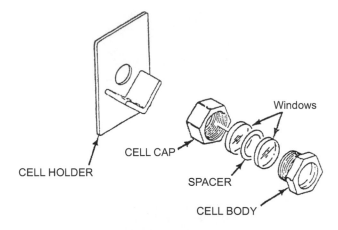

Figure 1.13 A typical infrared cell assembly for very volatile liquids.

then secured by the screw holder, placed in the cell holder and the spectrum obtained.

1.10.4 Solid Samples

The spectra of solid samples may be obtained as nujol mulls, KBr pellets or in solution. Nujol is selected as a mulling agent, since it is a mixture of high-molecular-weight hydrocarbons, exhibiting only a few absorptions bands resulting mainly from C–H stretching and bending vibrations (Fig. 1.14).

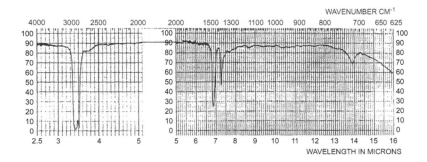

Figure 1.14 Nujol, liquid film.

In preparing a nujol mull, approximately 4–6 milligrams of the dry sample is ground to a fine powder, preferably using the smooth surface of an agate mortar and pestle. A drop of nujol (mineral oil) is added to the powered sample and the grinding continued until a thick paste is obtained. A thin film of the paste is spread between the two plates, and the spectrum obtained. The particle size of the finished sample and the thickness of the nujol mull between the plates are important factors in obtaining a satisfactory spectrum. If the film is too thick, light will not be adequately transmitted through the sample; if too thin, the spectrum will consist mostly of nujol bands. By rotating the salt plates against each other, a thick sample may be thinned sufficiently to give a satisfactory spectrum. Even a good spectrum will show absorptions due to the mulling agent. These absorptions, as shown in Fig. 1.14, occur in the region between 3030–2857 cm^{-1} (3.3–3.5 μm) and 1460–1374 cm^{-1} (6.85–7.28 μm), hence, any absorptions by the sample compound in these regions will be obscured. To compensate for this deficiency, it is customary to prepare a second mull, using either hexachloro-1,3-butadiene of Fluoroluble® (Fig. 1.15) as the mulling agent. Neither of these compounds absorbs in the nujol regions. Therefore, the

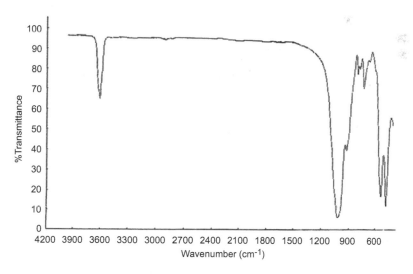

Figure 1.15 Fluorolube®, liquid film.

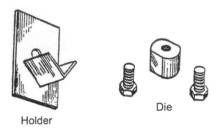

Holder

Die

Figure 1.16 A mini-press for preparing KBr pellets.

entire spectrum of the solid sample may be obtained using the two different mulling agents.

As mentioned, the spectra of solids may also be obtained as KBr pellets. This procedure involves grinding a mixture of about 1 milligram of the dry sample and approximately 100 milligrams of dry KBr. After grinding to a fine powder, the mixture is fused in a press. A common and inexpensive press consists of two stainless steel screws and a die (Fig. 1.16). Fusion is accomplished by placing the powdered mixture into the cavity of the die. Both screws are then manually tightened against the sample until a transparent solid solution is obtained. The screws are removed and the die containing the transparent solid is placed in the sample holder and attached to the infrared spectrometer. More elegant devices for preparing KBr pellets are commercially available. Some are capable of evacuating the sample chamber prior to fusion.

Because of the hygroscopic properties of KBr, it is difficult to completely exclude moisture from the finished KBr pellet. This is especially true if the fusion chamber is not evacuated. Thus, one should be aware that the presence of moisture could interfere with the recognition of N-H and O-H groups common to the sample. It should also be noted that amines may react with KBr during the fusion process. Lastly, heat generated during the fusion process may result in some decomposition of the sample compound. Any of these occurrences could complicate the analysis by producing extraneous absorption bands.

Occasionally, the spectrum of low melting solids may be obtained by heating the sample between salt plates. The melted sample (between the plates) is then placed in a holder and attached to the

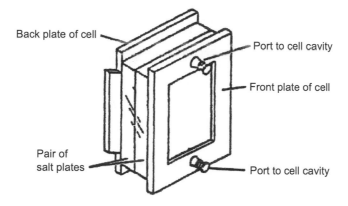

Figure 1.17 A typical solution cell.

infrared spectrometer. In order to obtain a satisfactory spectrum, the sample must remain a liquid until the spectrum is obtained. This method works best for samples melting between 20–30°C. Occasionally a melted sample may solidify as a transparent glassy melt, resulting in the transmission of infrared radiation and a good spectrum.

Infrared spectra of solids may also be obtained as 5–10% solutions. The pure solvent is first placed in a sodium chloride solution cell similar to the one shown in Fig. 1.17 with the aid of a syringe. The cell is sealed and attached to the sample beam of the instrument and the spectrum of the solvent obtained. The solvent is removed and the spectrum of the 5–10% solution is obtained in the same fashion. Most modern infrared spectrometers are designed to subtract the spectrum of the solvent from the solution leaving only the absorptions bands of the dissolved solid sample. However, in regions of the spectrum where the solvent absorbs strongly, the transmission of light will be nearly 100%. In such cases, not only will the absorptions bands of the solvent be absent, but any absorption by the sample in this region will also be absent. If carbon tetrachloride (Fig. 1.18) is the solvent, the region between 883–714 cm^{-1} (12–14 μm) will be totally devoid of both solvent and sample absorptions bands. For samples dissolved in chloroform (Fig. 1.19), the region lying near 1220 cm^{-1} (8.2 μm) and 781–714 cm^{-1} (12.8–14.0 μm) will be devoid of all absorptions.

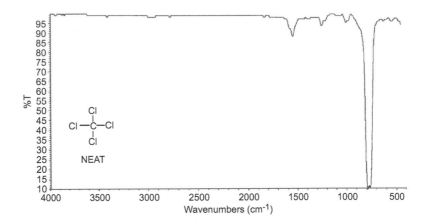

Figure 1.18   Carbon tetrachloride, liquid film.

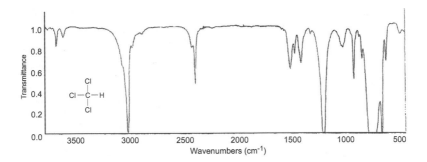

Figure 1.19   Chloroform, liquid film.

The entire spectrum of the sample compound may be realized by obtaining two spectra of the sample compound using different solvents. For instance, carbon tetrachloride, as stated, display strong absorption bands in the region between 883–714 cm^{-1} (12–14 μm), but for carbon disulfide (Fig. 1.20), this region of the infrared spectrum is relatively free of absorption bands. By obtaining spectra of the sample in both carbon disulfide and carbon tetrachloride, the complete spectrum of the sample compound may be studied.

If the material is a powered sample of small particle size, it is possible to obtain an acceptable spectrum by spreading a thin film

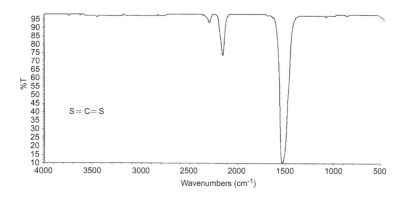

Figure 1.20   Carbon disulfide, liquid film.

of the powdered material on a salt plate. This procedure does not work if the particle size of the sample is relatively large.

Theoretically, the ideal infrared solvent for dissolving solid samples is one that is: (a) inert, (b) capable of dissolving a wide variety of organic compounds and (c) free of absorptions in the infrared region between 4000–625 cm^{-1} (2.5–16 μm). Unfortunately, no such solvent exists since they all show some absorption in the infrared region. As a compromise, one must settle for a solvent that will dissolve the sample compound, yet show only a few absorptions in the infrared region. In most cases, the solvent of choice is either carbon tetrachloride, chloroform of carbon disulfide (Figs. 1.18, 1.19, and 1.20, respectively). Of these three, the spectrum of chloroform has more absorptions bands because it is less symmetrical; however, it is generally the better solvent.

1.11 Other Sampling Techniques

In Section 1.10, basic infrared sampling techniques have been discussed. In addition to these techniques, there are others that allow the analyst to obtained excellent spectra on a variety of sample materials with little or no sample preparation. These materials include aqueous samples, pastes, gels, powders, films, fibers, and rigid plastics as well as micro samples. In most cases, these techniques require special infrared accessories. In Chapter 3 of this text, descriptions of the techniques, the accessories, and other considerations are discussed.

1.12 Some Suggestions and Comments on the Interpretation of Infrared Spectra

Naturally, the extent in which an infrared spectrum is analyzed depends upon the information desired. In some cases, all that may be required is a functional group verification. In other instances, it may be necessary to study the progress of a reaction by following the disappearance or appearance of a particular absorption. In still other instances, a more rigorous analysis of the spectrum may be necessary in order to acquire as much structural information as possible. The latter situation is more common when complimentary spectra data is limited or unavailable. Although there is no rigid approach to the interpretation of infrared spectra, some suggestions may be in order. These include the following.

Before obtaining infrared spectra, ensure to the extent possible that the sample is free from impurities and the spectrum is calibrated so that the absorptions appear at their proper positions. For modern infrared spectrometers, spectra calibration is automatic.

In analyzing an infrared spectrum, it is important to know the history of the sample and the medium that was used, i.e., if it is a nujol mull, KBr pellet, etc.

Remember that a structural correlation of all absorption bands in an infrared spectrum is unrealistic. There will always be absorptions in which structural correlations cannot be made.

Two important regions of an infrared spectrum lie between 4000–1299 cm^{-1} (2.5–7.7 µm) and 909–625 cm^{-1} (11.0–16 µm). It is the former region where stretching vibrations of most functional groups appear. Bending vibrations are mainly confined the latter region (Section 1.5).

The region of the infrared spectrum between 1300–909 cm^{-1} (7.7–11.0 µm) is referred to as the "fingerprint" region. This region is often characterized by a complex group of bending vibrations, which cannot always be correlated with the structure of the compound. Compounds of similar structural features may show almost identical absorptions in other regions of the spectrum, but will usually have their own unique group of bands in the "fingerprint" region. However, some groups of structurally similar compounds (such as the long chain hydrocarbons) may result in almost identical spectra.

Bending vibrations are usually more numerous and more difficult to correlate than stretching vibrations.

After verifying the presence of a particular functional group, it is important to seek out other diagnostic absorptions that will help support the class of compound.

If a particular functional group absorption is not present in the spectrum, it is usually safe to conclude that it is not present in the sample compound.

The position of an absorption may be shifted due to factors such as resonance, electron withdrawing and donating effects, steric interactions and hydrogen bonding. As you will see in Chapter 2, these shifts often convey important structural information about the sample compound.

The stretching vibrations of C–H, 2962–2853 cm^{-1} (3.38–3.51 µm); C=O, 1774–1667 cm^{-1} (5.6–6.0 µm) and O–H, 3650–3448 cm^{-1} (2.74–2.90 µm), may vary within a narrow range. Often, these small variations are of structural significance.

Unless hydrogen bonding is involved, the characteristic absorption of functional groups remains fairly constant in different solvents.

The exact position of an absorption is an important criterion in making structural assignments. Even if there is an extremely small shift of an absorption from its "normal" position, it may convey important structural information relating to electron withdrawing and donating effects, resonance, conjugation and other factors. In the analysis of spectra, the reader will ultimately memorize some common functional group absorptions. As result, general structural assignments may be made based on a cursory evaluation of the spectrum. For a more detailed analysis, correlation tables and charts (similar to the ones shown by Fig. 1.21, Table 1.4, and those listed in the appendix), will become indispensable.

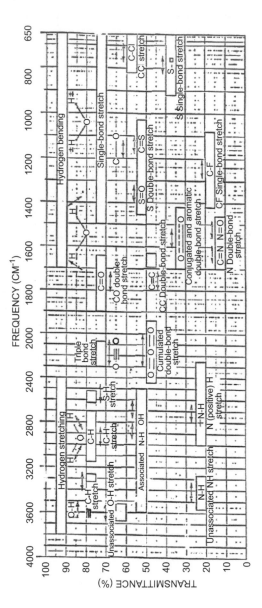

Figure 1.21 Absorptions in different regions of the infrared spectra. *Source:* Hannah RW, Swinchart JS, *Experiments in Techniques of Infrared Spectroscopy*, Perkin-Elmer Corp., 1967.

Table 1.4 Some important infrared absorption bands

Bond type	Class of compound	Wave number (cm^{-1})	Wavelength (μm)	Intensity*
C–H	Alkanes			
	–CH, two peaks	2960–2870	3.38 & 3.48	
	–CH$_2$, two peaks	2925 & 2855	3.42 & 3.50	
	Ring	3100–2990	3.23 & 3.34	
	Alkenes	Left of 3000	3.33	m–w
	Alkynes	3333–3267	3.00 & 3.06	m
	Aromatic	3100–3000	3.23–3.33	
	Aldehydes	2 peaks		m–w
		2900–2800	3.45–3.57	
		2830–2700	3.57–3.70	
C=C	Alkenes	1670–1640	5.99–6.10	m–w
	Conjugated dienes (symmetric)	1600	6.25	
	(unsymmetric)	1650 & 1600	6.06 & 6.25	
C to C	Aromatics	1600–1500 [1 peak]	6.25–6.67	m–w
	Ring	1500–1400 [1 or 2 peaks]	6.67–7.14	s–m
C≡C	Alkynes	2260–2100	4.24–4.76	m–w
C=O	Carboxylic acids	1720–1665	5.81–6.00	s
	Esters	1750–1715	5.71–5.83	s
	Amides	1700–1640	5.88–6.10	s
	Ketones	1725–1705	5.80–5.87	s
	conjugated to double bond	1710–1665	5.85–6.00	s
	Aldehydes	1740–1720	5.75–5.81	s

Table 1.4 *Continued*

Bond type	Class of compound	Wave number (cm^{-1})	Wavelength (μm)	Intensity*
C–O	Alcohols	1260–1000	7.94–10	s–m
	Acids	1315–1280 (doublet)	7.60–7.81	s–m
	Esters	1300–1000	7.69–10	s–m
	Aliphatic ethers	1150–1085	8.70–9.22	s–m
	Aromatic alkyl ethers	1275–1200	7.84–8.33	s–m
O–H	Alcohols, phenols	3649–3584	2.74–2.79	m
	(hydrogen bonded)	3550–3200	2.82–3.13	s
	Carboxylic acids	3300–2500	3.03–4.00	s
		(centered at 3000)		
N–H	Amines, primary	3500 & 3400	2.86 & 2.94	m
	secondary	3300–3060	2.99–3.02	
	Amides	3520–3060	2.84–3.27	s–m
C–N	Amines	1345–1020	7.43–9.80	m
	Amides	around 1400	around 7.14	s
C≡N	Nitriles	2260–2220	4.42–4.50	s
N=O	Nitro	2 peaks		
		1600–1500	6.25–6.67	s
		1400–1300	7.14–7.69	s

Source: Miller JA, Neuzil EF, *Modern Experimental Organic Chemistry*, D. C. Heath and Company, Massachusetts, 1980.

Note: *s = strong, m = medium, w = weak.

In Chapter 2, the analysis of infrared spectra will be discussed. In these discussions, efforts have been made to avoid repetition. For instance, if a C–H stretching vibration has been discussed under hydrocarbons, it is not discussed elsewhere, since its position and appearance, for the most part, remains fairly constant. In cases where the position or appearance of an absorption is different from one previously discussed and where this difference conveys important structural information, then discussion of the absorption is repeated in light of its structural significance. If carefully studied, Chapter 2 should give the reader the experience necessary to develop his or her own interpretive style.

In Chapter 2, you will observe infrared spectra obtained from both dispersive and Fourier transform infrared (FTIR) instruments. For FTIR spectra the absorption bands are linear with respect to wavenumbers (cm^{-1}), whereas in dispersive spectra, the band positions are linear with respect to wavelength. These differences will pose no problems since band positions do not change.

Chapter 2

The Analysis of Infrared Spectra

2.1 Hydrocarbons (Straight Chain)

Saturated straight chain hydrocarbons have relatively simple spectra (Fig. 2.1), showing mostly CH_2 and CH_3 stretching and bending vibrations.

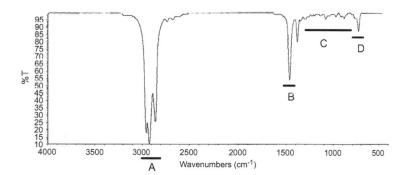

Figure 2.1 Octane, liquid film. (A) 2970–2850 cm^{-1} (3.37–3.51 μm): CH_2 and CH_3 stretching vibrations. (B) 1455 cm^{-1} (6.87 μm) and 1375–1385 cm^{-1} (7.00–7.22 μm): CH_2 and CH_3 bending vibrations. (C) "Fingerprint" region. (D) 725 cm^{-1} (13.80 μm): the concerted CH_2 rocking vibration.

Infrared Spectroscopy
James M. Thompson
Copyright © 2018 Pan Stanford Publishing Pte. Ltd.
ISBN 978-981-4774-78-9 (Hardcover), 978-1-351-20603-7 (eBook)
www.panstanford.com

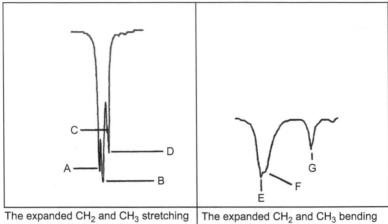

The expanded CH_2 and CH_3 stretching regions of straight chain hydrocarbons between 2970-2850 cm^{-1} (3.37-3.51 μ).	The expanded CH_2 and CH_3 bending regions of straight chain hydrocarbons between 1455 cm^{-1} (6.87 μ) and 1375-1385 cm^{-1} (7.00-7.22 μ).

Figure 2.2 The expanded CH_2 and CH_3 stretching and bending regions of straight chained hydrocarbons. (A) 2962 cm^{-1} (3.37 μm): asymmetrical stretching of the CH_3 group. This vibration is always found between 2952 and 2972 cm^{-1} (3.39–3.36 μm). (B) 2926 cm^{-1} (3.42 μm): asymmetrical H–C–H stretching vibration. This absorption normally occurs between 2916 and 2936 cm^{-1} (3.43–3.41 μm). (C) 2875 cm^{-1} (3.48 μm): symmetrical stretching of the CH_3 group. This absorption normally occurs in the 2872 ± 10 cm^{-1} range. (D) 2856 cm^{-1} (3.50 μm): symmetrical H–C–H stretching vibration. This absorption normally occurs in the 2853 ± 10 cm^{-1} range. (E) 1468 cm^{-1} (6.81 μm): asymmetrical bending vibration of the CH_3 group. This absorption usually occurs in the 1460 ± 10 cm^{-1} range. (F) 1445–1465 cm^{-1} (6.92–6.83 μm): the H–C–H scissoring bending vibration. (G) 1375–1385 cm^{-1} (7.00–7.22 μm): symmetrical bending vibration of the CH_3 group. This absorption is also known as the "umbrella" bending vibration since it reminds one of the opening and closing of an umbrella. This isolated absorption is characteristic of open chain hydrocarbons and is not found in cyclic alkanes (compare Figs. 2.1 and 2.5). *This feature is often used to distinguish the spectra of cyclic and non–cyclic hydrocarbons.*

As noted in the spectrum of octane (Fig. 2.1), the CH_2 and CH_3 stretching vibrations occur within a narrow range between 2970 and 2850 cm^{-1} (3.37–3.51 μm) and the major bending vibrations occur near 1455 cm^{-1} (6.87 μm) and 1375–1385 cm^{-1} (7.00–7.22 μm). The absorptions near 1455 cm^{-1} (6.87 μm) are usually characterized by peak overlapping (Fig. 2.2), due to both the CH_2 and CH_3 bending vibrations. The C–C stretching vibration is too weak to be of diagnostic value and the C–C bending vibration occurs near 500 cm^{-1} (20 μm) which is beyond the range of the spectrum. Expanded views of the CH_2 and CH_3 absorptions are shown in Fig. 2.2, followed by a more detailed analysis.

In Fig. 2.1, the absorption near 725 cm^{-1} (13.8 μm) is the concerted CH_2 rocking vibration. This absorption normally occurs in the 720 ± 10 cm^{-1} range. *The intensity of the absorption is often correlated with the number of consecutive CH_2 groups in the molecule.* If the number is less than six, there may be no absorption in this region, or it may be extremely weak. For hydrocarbons with six or more consecutive CH_2 groups, the absorption is usually of medium intensity. When observed in the solid state (nujol of KBr pellet), the band may be split.

2.2 Hydrocarbons (Branched Chain)

The spectra of branched and straight chain hydrocarbon are similar in many respects, yet there are several unique differences. For instance:

- The CH_3 umbrella bending vibration of isopropyl groups usually appears as a characteristic doublet of medium intensity, located between 1389 and 1379 cm^{-1} (7.20–7.25 μm), as shown in Fig. 2.3.
- Tertiary butyl groups are also characterized by a "doublet"; one member of the "doublet" occurs between 1395 and – 1380 cm^{-1} (7.17–7.25 μm) and the other near 1370 cm^{-1} (7.30 μm). The latter absorption is usually the more intense of the two (Fig. 2.4).
- If the sample compound has an internal carbon with the two methyl groups attached (a gem-dimethyl), the spectrum is also characterized by a "doublet" in the same region as the isopropyl and tertiary butyl "doublets" which appears near 1389–1379 cm^{-1} (7.20–7.25 μm).

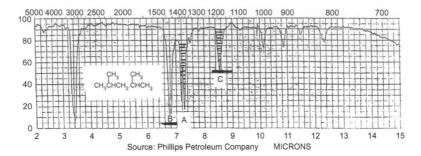

Source: Phillips Petroleum Company   MICRONS

Figure 2.3 2,4-Dimethylpentane, neat liquid. (A) 1388–1380 cm^{-1} (7.20–7.25 μm): the "isopropyl doublet." This absorption is due to the CH$_3$ umbrella bending vibration of the isopropyl groups. A "doublet" in the region between 1399 and 1370 cm^{-1} (7.15–7.30 μm) suggests an isopropyl, t-butyl or gem-dimethyl group. It should be emphasized that this band is not split into a "doublet" in the spectrum of octane (Fig. 2.1) and other nonbranched hydrocarbons. It should also be noted that the absorptions near 1399–1370 cm^{-1} (7.15–7.30 μm) is absent in the spectra of nonbranched cycloalkanes (Figs. 2.5a and 2.5b). (B) 1470 cm^{-1} (6.8 μm): asymmetrical bending vibration of the CH$_3$ group. (C) 1172 cm^{-1} (8.53 μm), skeletal vibration, a characteristic of branched hydrocarbons.

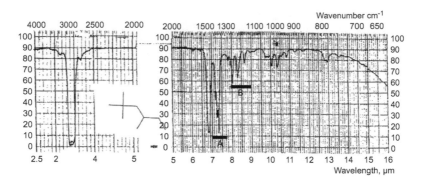

Figure 2.4 2,2,4-Trimethylhexane, liquid film. *Source*: The Aldrich Library of Infrared Spectra © Aldrich Chemical Company, Inc. (A) 1388–1374 cm^{-1} (7.20–7.25 μm): the tertiary butyl doublet. Note: the member of the doublet at higher wavelength has the greater intensity, a characteristic feature of gem-dimethyl groups. (B) 1250–1149 cm^{-1} (8.0–8.7 μm): skeletal vibrations, a characteristic of branched hydrocarbons.

- The spectra of branched hydrocarbons may show several absorptions bands between 1520 and 910 cm^{-1} (8–11 μm) due to skeletal vibrations (Figs. 2.3 and 2.4).

2.3 Hydrocarbons (Cyclic)

Unless there is ring strain, the spectra of cyclic hydrocarbons are similar to those of straight chain hydrocarbons. There is, however, one major exception. *Non-branched cyclic alkanes do not show absorptions in the 1385–1365 cm^{-1} (7.22–7.33 μm) region of the spectrum.* In the discussion of the spectrum of octane (Fig. 2.1), it was noted that this absorption is due to the symmetrical bending vibration ("umbrella" bend) of the CH$_3$ group. Also, for cyclic alkanes there is a very slight shift of the CH$_2$ bending vibration to lower wavenumber (higher wavelength). Compare, for instance, the combined spectra of octane and cyclohexane (Fig. 2.5). In octane the CH$_2$ bending vibration appears near 1468 cm^{-1} (6.81 μm), while in cyclohexane, it is located near 1461 cm^{-1} (6.88 μm).

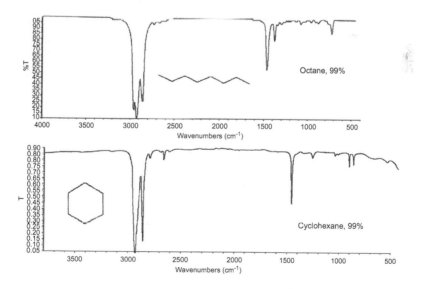

Figure 2.5 The combined spectra of octane and cyclohexane.

2.4 Alkenes

If the alkene contains a terminal double bond, there is usually a sharp band near 3125 cm^{-1} (3.2 µm), which is due to the CH_2 olefinic stretching vibration (Fig. 2.6). As a general rule, if a CH stretching absorption occurs above 3000 cm^{-1} (below 3.33 µm), there is a good chance that some type of unsaturation exists in the compound (double bond, triple bond or aromatic ring). In general, the CH_2 olefinic stretching vibration appears slightly above 3000 cm^{-1} (3.33 µm) as a distinct band; however, occasionally this absorption may be obscured by other C–H stretching vibrations in the vicinity (compare Figs. 2.6, 2.7, and 2.8 with Fig. 2.9).

Equally as characteristic of the spectra of alkenes is the C=C stretching vibration which appears as a sharp but weak to medium absorption between 1667 and 1613 cm^{-1} (6.0–6.2 µm), as shown in Figs. 2.6, 2.7, 2.8, and 2.9. If the alkene is vinylic or *cis*, the C=C

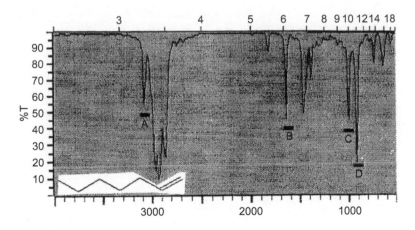

Figure 2.6 1-Heptene, liquid film. (A) 3096 cm^{-1} (3.23 µm): the sharp absorption represents the olefinic =CH_2 stretching vibrations of the terminal double bond. (B) 1631 cm^{-1} (6.13 µm): C=C stretching vibrations. (C) 990 cm^{-1} (10.1 µm): The vinyl (R–CH=CH_2) deformation. This absorption represents the =CH_2 out-of-plane twist. It is observed only in alkenes having *trans* olefinic hydrogens. For *cis* olefinic hydrogens there is a different out-of-plane bending absorption. (D) 903 cm^{-1} (11.07 µm): =CH_2 out-of-plane bending. This absorption appears only in compounds with terminal double bonds (compare Figs. 2.6, 2.7, and 2.8 with Fig. 2.9).

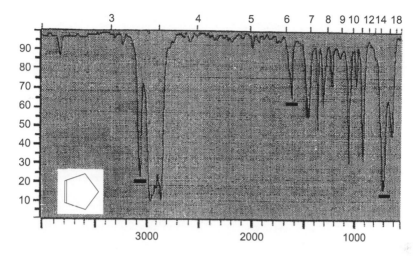

Figure 2.7 Cyclopentene, liquid film.

stretching vibration appears between 1620 and 1660 cm^{-1} (6.17–6.02 μm). If the alkene is *trans*, tri or tetra substituted, the C=C stretching occurs between 1660 and 1680 cm^{-1} (6.02–5.95 μm).

In Fig. 2.6, the –CH=CH$_2$ vinyl deformations appear as two relative strong absorptions near 1000 and 910 cm^{-1} (10 and 11 μm). If one of the carbon of the terminal double bond is substituted as in RR'C=CH$_2$, the absorption near 1000 cm^{-1} (10 μm) disappears

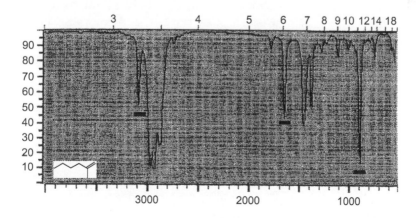

Figure 2.8 2-Methyl-1-hexene, liquid film.

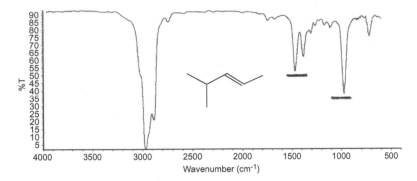

Figure 2.9 *trans*-4-Methyl-2-pentene, liquid film. (A) 967 cm^{-1} (10.34 μm): C–H olefinic out-of-plane bending vibration. Note the absence of the =CH$_2$ out-of-plane bending near 900 cm^{-1} (11.11 μm). As mentioned, this absorption appears only in those compounds containing terminal double bonds (compare Fig. 2.6 and 2.8 with Fig. 2.9). (B) Note the isopropyl "doublet" near 1389 cm^{-1} (7.27 μm) and the presence of a relatively weak C=C stretching vibration near 1631 cm^{-1} (6.13 μm).

(compare Figs. 2.6 and 2.8). For alkenes such as RR′C=CRH, the absorption near 910 cm^{-1} (11 μm) is absent or shifted to around 833 cm^{-1} (12 μm).

In the spectrum of cyclopentene (Fig. 2.7), the broad absorption centered near 699 cm^{-1} (14.3 μm) is characteristic of *cis* orientation of cycloalkenes. If the olefin is disubstituted, fairly symmetrical, and the double bond is internally located, the C=C stretching absorption between 1667 and 1631 cm^{-1} (6.0–6.2 μm) is absent or very weak, since there is little or no change in the dipole moment during vibration. Because of greater molecular symmetry, the C=C stretching absorptions of *trans* olefins are usually weaker than they are for the *cis* olefins. Conjugated, unsymmetrical olefins may show two C=C stretching vibrations near 1650 cm^{-1} (6.06 μm) and 1600 cm^{-1} (6.25 μm). If conjugated with an aromatic ring, the stretching vibration of the C=C double bond is shifted and appears near 1625 cm^{-1} (6.15 μm). If conjugated with a carbonyl group, the C=C stretching vibration is usually lowered by approximately 30 cm^{-1}. Frequently, the intensity of the C=C stretching band is increased with conjugation.

The position of the double bond in the spectra of cycloalkenes is influenced by ring strain. The greater the degree of strain, the lower the frequency (higher the wavelength) of the C=C stretching vibration. This point is illustrated in the examples below:

Some C=C stretching vibrations for cycloalkenes

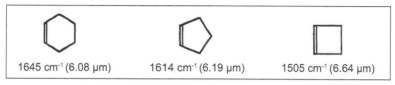

| 1645 cm⁻¹ (6.08 µm) | 1614 cm⁻¹ (6.19 µm) | 1505 cm⁻¹ (6.64 µm) |

2.5 Alkynes

The normal range for the acetylenic (≡C–H) stretching vibrations for terminal triple bonds is between 3320 and 3280 cm^{-1} (3.01–3.05 µm). As shown in Figs. 2.10 and 2.11, this absorption is usually strong and sharp. The C≡C stretching vibration of the terminal triple bond appears between 2260 and 2100 cm^{-1} (4.43–4.76 µm) as shown in Figs. 2.10 and 2.11. If both carbons of the triple bond are substituted, the C≡C stretching vibration will appear between 2215

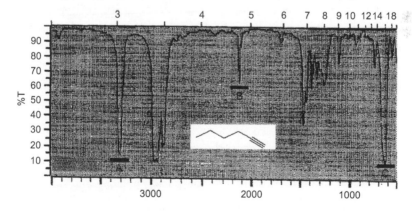

Figure 2.10 1-Hexyne, liquid film. (A) 3280 cm^{-1} (3.04 µm): ≡C–H acetylenic stretching vibration. (B) 2119 cm^{-1} (4.72 µm): C≡C stretching vibration. (C) Centered near 667 cm^{-1} (near 15 µm): acetylenic ≡CH wagging vibration.

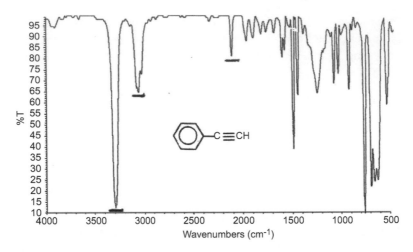

Figure 2.11 Phenyl acetylene. (A) 3300 cm^{-1} (3.03 μm): ≡C–H acetylenic stretching vibration. (B) 2119 cm^{-1} (4.72 μm): C≡C stretching vibration. (C) Note the C–H aromatic stretching vibration near 3077 cm^{-1} (3.25 μm).

and 2235 cm^{-1} (4.51–4.47 μm). Symmetrical of nearly symmetrical alkynes show little or no dipole moment change during vibration, consequently, the C≡C stretching vibration is either weak or absent. The spectra of terminal alkynes also exhibit broad ≡CH wagging vibration in the region between 665 and 625 cm^{-1} (15–16 μm). Figures 2.10 and 2.11 show examples of such an absorption.

2.6 Ethers (Alkyl–Alkyl)

As expected, the spectra of straight chain aliphatic ethers and straight chain aliphatic hydrocarbons are similar with the exception of a broad and strong C–O stretching vibration between 1149 and 1087 cm^{-1} (8.7–9.2 μm). In comparing alkanes and aliphatic ethers, consider the spectra of octane and butyl ether (Figs. 2.12 and 2.13).

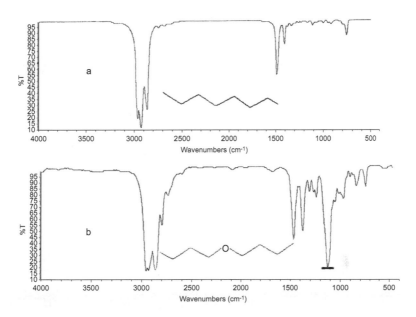

Figure 2.12 Combined spectra of octane (a) and butyl ether (b), liquid films.

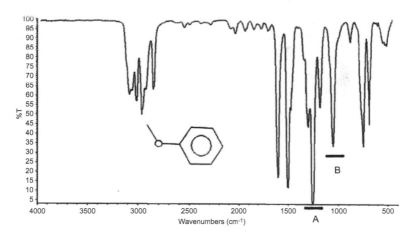

Figure 2.13 Methylphenyl ether, neat. (A) 1242 cm^{-1} (8.05 μm): C–O aryl stretching vibration. (B) 1042 cm^{-1} (9.6 μm): The C–O alkyl stretching vibration.

2.7 Ethers (Aryl–Alkyl)

In aryl-alkyl ethers, both the aryl and alkyl C–O stretching vibrations appear as strong absorptions, but at different positions (Figs. 2.13 and 2.14). The aryl C–O stretching vibration occurs near 1250 cm^{-1} (8.0 µm) and the alkyl C–O stretching absorption appears between 1053 and 1000 cm^{-1} (9.5-10.0 µm). The spectrum of methyl phenyl ether (Fig. 2.13) shows both the aryl and alkyl C–O absorptions. These absorptions are summarized below and may be used to distinguish among alkyl-alkyl, aryl-alkyl and aryl-aryl ethers.

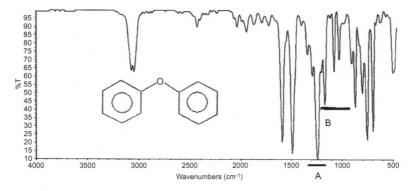

Figure 2.14 Phenyl ether. (A) 1242 cm^{-1} (8.05 µm): C–O aryl stretching vibration. Note the absence of the strong C–O alkyl stretching vibration which normally occurs between 1053 and 1000 cm^{-1} (9.5-10 µm). (B) 1175-1020 cm^{-1} (8.5-9.8 µm): aromatic ring vibrations.

Some C–O stretching vibrations of ethers

Class of ether	Aryl C–O	Alkyl C–O
Aryl–Aryl	Near 1250 cm^{-1} (8.0 µm)	None
Aryl–Alkyl	Near 1250 cm^{-1} (8.0 µm)	1053–1000 cm^{-1} (9.5–10.0 µm)
Alkyl-Alkyl	None	1053–1000 cm^{-1} (9.5–10.0 µm)

2.8 Aliphatic Halides

An outstanding spectral feature of alkyl chlorides and fluorides is the strong and broad C–X stretching vibration. The C–I and C–Br stretching occurs beyond the infrared range of some instruments as shown below. The strong and broad C–Cl stretching vibration is illustrated by the spectra of carbon tetrachloride (Fig. 1.18) and chloroform (Fig. 1.19). Figure 1.15 shows the C–F stretching of Fluorolube®.

Some C–X stretching vibrations of alkyl halides

Group	Wavenumbers (cm^{-1})	μm (microns)	Intensity
C–F	1400–1000	7.1–10.0	Strong
C–Cl	800–600	12.5–16.6	Strong
C–Br	600–500	16.6–20	Strong
C–I	500	20.0	Strong

2.9 Amines

Primary amines are characterized by absorptions in the following general regions.

Two weak absorptions attributed to the $-NH_2$ asymmetrical and symmetrical stretching vibrations. These absorptions are located near 3390 cm^{-1} (2.95 μm) and 3296 cm^{-1} (3.03 μm) respectively. The normal range of the asymmetrical stretching is between 3400 and 3200 cm^{-1} (2.94–3.13 μm). In secondary amines, there is only one absorption in the regions mentioned and for tertiary amines there are none.

$-NH_2$ scissoring bending vibrations: near 1615 cm^{-1} (6.2 μm). The normal range for this absorption is 1600–1630 cm^{-1} (6.25–6.13 μm). Obviously, this absorption is absence in the spectra of secondary and tertiary amines.

$-NH_2$ wagging: between 910 and 770 cm^{-1} (11 and 13 μm). This broad absorption is usually found in the spectra of secondary amines, but is not found in the spectra of tertiary amines.

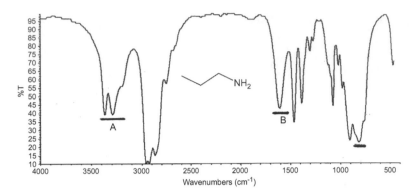

Figure 2.15 Propylamine, liquid film. (A) 3335 cm^{-1} (3.00 μm): N–H stretching vibration (note the "split" absorption which is unique to the spectra of primary amines, when taken as a neat liquid). For secondary amines only one absorption is seen in this region and tertiary amines show none. (B) 1613 cm^{-1} (6.20 μm): NH$_2$ scissoring vibration (not detected in secondary and tertiary amines (compare Figs. 2.15 and 2.16). (C) The broad band centered near 800 cm^{-1} (12.50 μm) is the N–H wagging vibration. The position and intensity of this band depends upon the extent of hydrogen bonding. (D) Again, the spectra of secondary amines, whether in solution or taken neat show only one N–H stretching vibration which is located near 3335 cm^{-1} (3.0 μm). This band is in the same general region as the N–H stretching vibration of primary amines (compare Figs. 2.15 and 2.16).

The appearance of the NH$_2$ stretching vibrations for primary amines is dependent upon the degree of hydrogen bonding. In solution, the NH$_2$ stretching vibration has the appearance of a weak doublet resulting from asymmetrical and symmetrical stretching. When the spectrum is of the neat liquid, hydrogen bonding is more extensive and the two absorptions often coalesce to give the appearance of a split band (Fig. 2.15) or a singlet, near 3333 cm^{-1} (3 μm). For secondary amines, the absorption near 3333 cm^{-1} (3 μm) appears as a singlet.

The NH$_2$ bending vibrations of primary amines appear near 1615 cm^{-1} (6.2 μm) as weak absorptions. Although located in different regions of the spectrum, the appearance and intensity of this absorption is often similar to the N–H stretching vibration (Fig. 2.15). Another prominent band in the spectra of primary amines is the broad and strong NH$_2$ wagging vibration (Fig. 2.15),

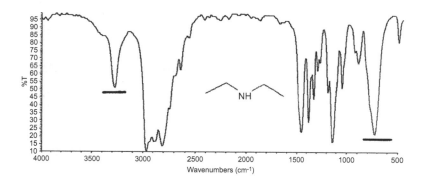

Figure 2.16   Diethylamine, liquid film.

which appears between 910 and 770 cm^{-1} (11 and 13 μm). In the absence of complicating factors, one can often distinguish among primary, secondary and tertiary amines.

The N–H bending vibration of primary amines, which normally appears near 1615 cm^{-1} (6.2 μm), is absent in secondary and tertiary amines (compare Figs. 2.15 and 2.16). The broad and strong absorption near 800 cm^{-1} (12.50 μm), which is attributed to NH$_2$ wagging vibrations of primary amines, is shifted to near 715 cm^{-1} (14 μm) in secondary amines (compare Figs. 2.15 and 2.16). Tertiary amines, as expected, show no N–H stretching, bending or wagging vibration. Although the C–O and C–N stretching vibrations usually appear in the same general region as the N–H wagging vibrations, the C–N stretching vibration is usually weaker and more dependent upon the structural features of the amine as shown below.

Some C–N Stretching Vibrations of Amines

General Structure	The C–N Stretching Vibration
RCH$_2$–NH$_2$	Near 1065 cm^{-1} (9.4 μ)
RCH$_2$–NHR	Near 1110 cm^{-1} (9.0 μ)
R$_2$CH–NH$_2$	Near 1150 cm^{-1} (8.7 μ)
R$_3$C–NH$_2$	Near 1235 cm^{-1} (8.1 μ)

Many of the band positions of amines, taken in the solid state (nujol mulls and KBr pellets) are not always reliable, since they are often split and shifted from their expected positions. The reader

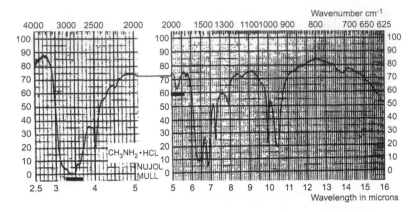

Figure 2.17 Methylamine hydrochloride, nujol mull. *Source*: The Aldrich Library of Infrared Spectra ⓒ Aldrich Chemical Company, Inc.

should also be reminded that amines are generally hygroscopic compounds, thus what may appear to be N–H stretching and bending vibrations, may instead be the O–H stretching and bending vibrations of water. For this reason, any absorption near 3333 cm^{-1} (3.0 μm) and 1615 cm^{-1} (6.2 μm) should be examined very closely.

The NH$_2$ stretching vibrations of primary amines salts are shifted from around 3333 cm^{-1} (3 μm) in the "free" amine to between 3030 and 2630 cm^{-1} (3.3–3.8 μm). These absorptions are usually much broader and stronger that they are for "free" amines (compare Figs. 2.15 and 2.17). The N–H bending vibrations of amines salts appear near 1429 cm^{-1} (7.00 μm), also as strong and broad absorptions. For primacy amine salts there is often a characteristic band between 2220 and 1820 cm^{-1} (4.5–5.5 μm). This band is usually not observed in secondary and tertiary amine salts.

Secondary amine salts, like primary amine salts, show strong absorptions in the region between 3000 and 2700 cm^{-1} (3.33 and 3.70 μm).

2.10 Ketones and Aldehydes

Ketones and aldehydes exhibit strong and conspicuous carbonyl stretching vibrations in the region between 1740 and 1665 cm^{-1}

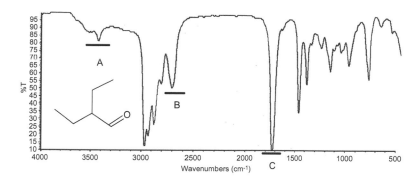

Figure 2.18 2-Ethylbutyraldehyde, liquid film. (A) 3448 cm^{-1} (2.9 μm): Overtone band of the C=O stretching vibration at 1724 cm^{-1} (5.80 μm). Note: the overtone band appears at a frequency which is twice the C=O frequency at 1724 cm^{-1} (5.80 μm). (B) 2825 cm^{-1} (3.54 μm) and 2714 cm^{-1} (3.68 μm): the weak C–H aldehyde stretching vibrations. This latter absorption appears between 2714 and 2750 cm^{-1} (3.68–3.64 μm) and is seldom obscured by the C–H stretching vibration. The absorption is useful in identifying aldehydes but should be used in combination with the C=O stretching vibration. (C) 1724 cm^{-1} (5.80 μm): C=O stretching vibration.

(5.75–6.00 μm), as shown by the spectra in Figs. 1.9, 2.18, and 2.19. The position of the carbonyl stretching vibrations is influence by conjugation, steric and electron-withdrawing and -donating effects.

Aldehydes generally exhibit two weak C H aldehyde stretching vibrations near 2857 and 2740 cm^{-1} (3.50 and 3.65 μm), as shown in Fig. 2.18; however, these absorptions are often obscured by C–H aliphatic stretching vibrations. This is especially true for the

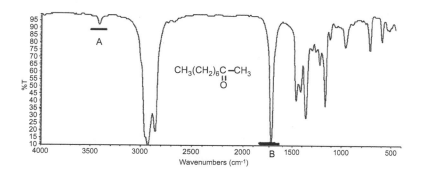

Figure 2.19 2-Nonanone, liquid film.

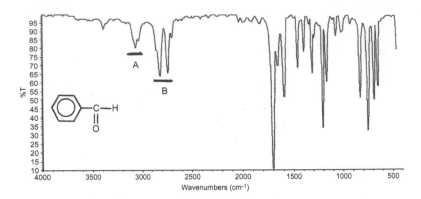

Figure 2.20 Benzaldehyde, liquid film. (A) 3125 cm^{-1} (3.25 μm): aromatic C–H stretching vibration. (B) 2833 and 2755 cm^{-1} (3.53 and 3.63 μm): aldehyde C–H stretching vibrations (asymmetrical and symmetrical).

absorption near 2857 cm^{-1} (3.50 μm). The absorption near 2740 cm^{-1} (3.65 μm) is usually not obscured by other absorptions (Fig. 2.18) and is therefore a good characteristic absorption for aldehydes, but it should be used in combination with the C=O stretching vibration. With aromatic aldehydes, containing no aliphatic hydrogens, observation of both of the two weak C–H aldehyde stretching vibrations is usually not a problem (Fig. 2.20). This is apparent since the C–H aromatic stretching vibrations occur near 3333 cm^{-1} (3.0 μm).

The various classes of carbonyl compounds, in addition to having their own unique spectral characteristics, can often be distinguished from each other by a precise measurement of their C=O stretching absorptions (Table 2.1).

As previously mentioned, the vibrational frequencies of bonds in organic compounds are influenced by electron-withdrawing and -donating effects. This may be illustrated by a consideration of the carbonyl stretching vibrations of *p*-substituted aromatic aldehydes (Table 2.2).

As observed in Table 2.3, there is a slight increase in the stretching frequency of the carbonyl group with increasing electron-withdrawing properties of the substituent. This trend may be explained by a consideration of the effect that resonance has upon the carbonyl double bond. In the case of electron donating

Table 2.1 Some C=O stretching vibrations for carbonyl compounds*

Potassium acetate	Salts of carboxylic acids	1575	6.35
N,N-Dimethylacetamide	Amides (3°) of carboxylic acids	1667	6.00
N-Methylacetamide	Amides (2°) of carboxylic acids	1669	5.99
Acetamide	Amides (1°) of carboxylic acids	1684	5.94
		1718	5.84
		(shoulder)	(shoulder)
Ethyl carbamate	Carbamates	1689	5.92
Butyric Acid (dimer)	Carboxylic acids (dimers)	1721	5.81
Butanone	Ketones	1724	5.80
Propyl formate	Esters of formic acid	1733	5.77
Butanal	Aldehydes	1736	5.76
Ethyl propionate	Esters of carboxylic acids	1736	5.76
Methyl propionate	Methyl esters of carboxylic acids	1748	5.72
Diethyl carbonate	Normal carbonates	1751	5.71
Diphenyl carbonate	Vinyl-type carbonates	1761	5.68
Isopropylidene acetate	Vinyl esters of carboxylic acids	1764	5.67
Vinyl acetate	Vinyl esters of carboxylic acids	1770	5.65
Phenyl acetate	Phenyl esters	1770	5.65
Butyric acid (monomer)	Carboxylic acids (monomers)	1776	5.63
Acetyl chloride	Chlorides of carboxylic acids	1812	5.52
Acetic anhydride	Anhydrides of carboxylic acids	1825	5.48

Source: Freeman, S. K. (editor), *Interpretive Spectroscopy*, Chapter 2, New York, Reinhold Publishing Company (1965). Small variations must be expected for various members of the same class.

*All values determined in carbon tetrachloride expect potassium acetate and ethyl carbamate (KBr pellet) and acetamide (chloroform).

Table 2.2 The influence of electron donating and withdrawing groups on the carbonyl stretching vibrations of aromatic aldehydes

X	C=O stretching (cm^{-1})	Microns (μm)
$-OC_2H_5$	1669 cm^{-1}	5.99 μm
$-CH_3$	1689 cm^{-1}	5.92 μm
$-Cl$	1701 cm^{-1}	5.88 μm
$-NO_2$	1703 cm^{-1}	5.87 μm

groups such as alkyl and alkoxy groups, the resonance effect, as shown below, will operate to decrease the double bond character of the carbonyl group. This results in a lower stretching frequency and a corresponding absorption at higher wavelengths. Electron-withdrawing groups exert an opposite inductive effect, which serves to maintain or increase the double bond character of the carbonyl group, leading to a higher stretching frequency and absorption at shorter wavelength. Both effects are illustrated below.

Based upon the above explanation, one would expect the carbonyl stretching frequency of *p*-ethoxybenzaldehyde to be greater than that of *p*-chlorobenzaldehyde (since the electronegativity of oxygen is greater than chlorine); however, this is not the case. A plausible explanation for what appears to be a contradiction lies in an understanding of the contribution that the lone pair of electrons on oxygen makes to the resonance hybrid of *p*-ethoxybenzaldehyde. As illustrated below, the non-bonded electrons of oxygen engage in resonance with the aromatic ring, producing resonance contributions which decrease the double bond character of the carbonyl group. The result is a stretching frequency lower that what would be expected, based on electron-withdrawing and -donating effects alone.

When the carbonyl group of aldehydes and ketones are conjugated with double bonds, the result is a shift of the absorption to lower frequencies. An example of this shift is shown below.

1715 cm⁻¹ (5.83 µm)　　　1718 cm⁻¹ (5.82 µm)　　　1682 cm⁻¹ (5.95 µm)

1632 cm⁻¹ (6.13 µm)　　　　1660 cm⁻¹ (6.02 µm)

The decrease in frequency of the carbonyl absorption with conjugation can be explained on the basis of resonance. As shown below, resonance interaction decreases the double bond character of the conjugated carbonyl group, resulting in an absorption at lower frequency.

$$RCH_2CH{=}CH{-}\overset{\|}{\underset{O}{C}}{-}R'$$

$$RCH_2\overset{+}{C}H{-}CH{=}C{-}R \\ \qquad\qquad\underset{O}{|}$$

A somewhat remarkable example of the shift in the stretching vibration of conjugated carbonyls is seen below by comparing the C=O absorption of 2,4,6-cycloheptatrienone and cycloheptanone.

1650 cm⁻¹ (6.06 µm)　　　　　　1695 cm⁻¹ (5.90 µm)

As a result of ring strain, cyclic ketones with less than six ring carbons show a slight shift in the C=O stretching vibration toward higher frequency (compare Figs. 2.19, 2.20, and 2.22).

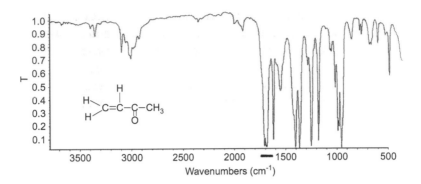

Figure 2.21   3-Butene-2-one.

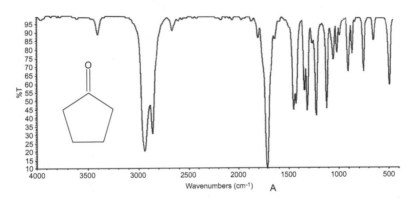

Figure 2.22   Cyclopentanone, liquid film. (A) 1738 cm^{-1} (5.75 μm) the carbonyl stretching vibration showing the slight shift toward higher frequency as a result of ring strain. (B) Compare the above carbonyl absorption with that of 2-nonanone (Fig. 2.19), which appears near 1709 cm^{-1} (5.85 μm).

2.11 Alcohols

The strong and broad O–H stretching vibration is the most prominent absorption of alcohols. For simple alcohols, the appearance and position of this absorption is influence by the degree of hydrogen bonding. If the spectrum is of the neat alcohol, hydrogen bonding is extensive and the O–H absorption appears between 3500 and 3200 cm^{-1} (3.13–2.75 μm) as a broad band (Figs. 1.8, 2.22, and 2.24). In dilute solution, hydrogen bonding is not as extensive and the absorption band is sharper and appears between 3640 and 3610 cm^{-1} (2.75 and 2.77 μm). If the O–H group is bonded in an intramolecular fashion (as shown below), the position of the O–H stretching remains fairly constant on dilution. When hydrogen bonding is intermolecular, dilution results in a sharper absorption band as well as a distinct shift to slightly shorter wavelength.

Another characteristic absorption of alcohols is the C–O stretching vibration. As shown in Figs. 1.8, 2.23, 2.24, and 2.25, this band also appears as a broad and strong absorption between 1260 and 1000 cm^{-1} (7.39–10.0 μm). *When taken neat, the spectrum of simple primary, secondary and tertiary alcohols may be distinguished by the positions of their C–O stretching absorptions.* For instance, the C–O stretching vibrations of primary, secondary and tertiary alcohols vary as follows:

Primary alcohols: 1075–1025 cm^{-1} (9.30–9.76 μm)
Secondary alcohols: 1000–1150 cm^{-1} (9.09–8.69 μm)
Tertiary alcohols: 1200–1100 cm^{-1} (9.30–9.76 μm)

For a more detailed assessment of the variations in the C–O stretching vibrations of alcohols, consider Table 2.3. In addition, a comparison of Figs. 2.23, 2.24, and 2.25 will reveal an obvious shift of the C–O stretching vibration as one proceed from primary through secondary to tertiary alcohol.

The O–H bending vibration for alcohols occurs in the region between 1420 and 1330 cm^{-1} (7.04–7.52 μm). This absorption often overlaps with the CH$_2$ bending vibration.

Table 2.3 Classification of some alcohols based on their C–O stretching vibrations*

Group	Position of C–O bands	Type of alcohol
1	1205–1124 cm^{-1} (8.30–8.90 μ)	(a) Saturated tertiary
		(b) Highly symmetrical secondary
2	1124–1087 cm^{-1} (8.90–9.20 μ)	(a) Saturated secondary
		(b) α-Unsaturated or cyclic tertiary
3	1086–1050 cm^{-1} (9.22–9.52 μ)	(a) α-Unsaturated secondary
		(b) Alicyclic secondary (5- or 6-membered ring)
		(c) Saturated primary
4	<1050 cm^{-1} (>9.52 μ)	(a) Highly α-unsaturated tertiary
		(b) Di-α-unsaturated secondary
		(c) α-unsaturated and α-branched secondary
		(d) Alicyclic secondary (7- or 8-membered ring)
		(e) α-Branched and/or α-unsaturated primary

Source: Zess, H. H., and M. Tsutsui, *J. Am. Chem. Soc.*, 75, 897 (1953).
*The above values are valid for pure liquids but must be increased slightly for spectra obtained in solution.

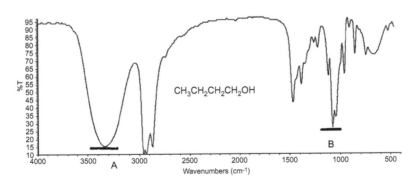

Figure 2.23 n-Butyl alcohol, liquid film. (A) Centered near 3300 cm^{-1} (3.03 μm): the O–H stretching vibration. (B) Centered near 1045 cm^{-1} (9.57 μm): the C–O stretching vibration. (C) Centered near 1400 cm^{-1} (7.14 μm): O–H bending deformation which is obscured by the CH bending vibrations.

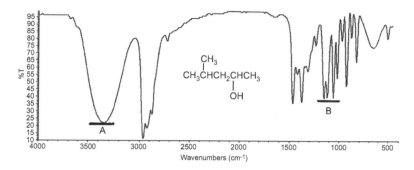

Figure 2.24 4-Methyl-2-pentanol, liquid film. (A) 3350 cm^{-1} (3.0 μm): the O–H stretching vibration. (B) Centered near 1136 cm^{-1} (8.8 μm): the C–O stretching vibration.

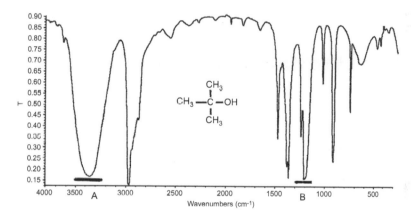

Figure 2.25 *tert*-Butyl alcohol, liquid film. (A) 3333 cm^{-1} (3.0 μm): the O–H stretching vibration. (B) Centered near 1190 cm^{-1} (8.4 μm): the C–O stretching vibration. (C) Note the position of the C=O stretching vibrations of the primary alcohol (Fig. 2.23), the secondary alcohol (Fig. 2.24) and the tertiary alcohol (Fig. 2.25). As noted, the position of these bands may be used to distinguish among primary, secondary, and tertiary alcohols.

2.12 Organic Acids

Organic acids show characteristic stretching absorptions for the O–H, C–O and C=O groups. In the solid state, neat and in solution (unless the solution is very dilute) carboxylic acids undergo hydrogen bonding to form dimers (see below). As a result, the

O–H stretching vibration appears as a broad and strong absorption in the region between 3300 and 2500 cm^{-1} (3.03–4.00 μm) as shown in Figs. 1.10 and 2.26.

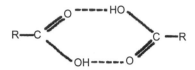

Generally, the carbonyl stretching vibrations of acids are more intense than they are for ketones or aldehydes. The absorption appears between 1725 and 1695 cm^{-1} (5.8–5.9 μm). In cases where the acid carbonyl is conjugated with a double bond, its absorption is shifted slightly to lower frequency, appearing between 1710 and 1680 cm^{-1} (5.85–5.95 μm). The shift of the carbonyl stretching vibration in carboxylic acids as a result of conjugation is less than it is for conjugated ketones and aldehydes.

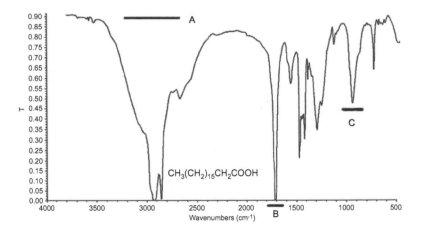

Figure 2.26 Stearic acid. (A) Near 2941 cm^{-1} (3.4 μm): broad and strong absorption of the O–H stretching vibration (shaded). Note the CH$_2$ stretching near 2898 cm^{-1} (3.45 μm) which overlaps with the O–H stretching vibration. (B) 1681 cm^{-1} (5.95 μm): C=O stretching vibration. (C) 939 cm^{-1} (10.65 μm): O–H out-of-plane bending vibration. (D) The C–O stretching vibration is shown near 1289 cm^{-1} (7.7 μm). This absorption is probably a combination of C–C stretching, O–H bending and C–O stretching occurring in unison.

The O–H bending vibrations of carboxylic acids appear between 950 and 910 cm^{-1} (10.5–11.0 μm) as a broad and medium absorption (Fig. 2.26).

2.13 Acid Halides

Because of the electron-withdrawing effects of the halogen, the carbonyl stretching vibrations of acid halides appear between 1815 and 1785 cm^{-1} (5.51–5.60 μm). Compare this value with the C=O stretching vibration of unconjugated aliphatic acids (Figs. 1.10 and 2.26) which appear between 1720 and 1685 cm^{-1} (5.81–5.90 μm). Usually when the alpha carbon of an acid halide is substituted, the carbonyl absorption is split (Fig. 2.27). A broad and strong absorption found between 1000 and 910 cm^{-1} (10–11 μm) is another characteristic absorption of acid halides (Fig. 2.27).

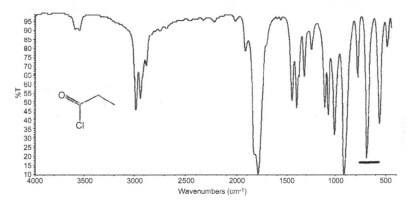

Figure 2.27 Propionyl chloride, neat. (A) 1786 cm^{-1} (5.6 μm): C=O stretching vibration. (B) Near 690 cm^{-1} (14.5 μm): the C–Cl stretching vibration.

2.14 Amides

Characteristic spectral features of amides include the N–H and C=O stretching vibrations (Figs. 2.28 and 2.29). For primary and secondary amides, the N–H bending vibration usually overlaps

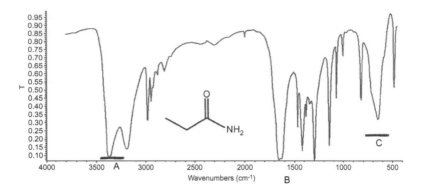

Figure 2.28 Propionamide, nujol mull. (A) 3226 cm^{-1} (3.1 μm): NH$_2$ stretching vibrations. (B) 1613 cm^{-1} (6.2 μm): carbonyl stretching vibration (the broad absorption is due to the overlap of the carbonyl group with the NH$_2$ bending vibration). (C) Near 667 cm^{-1} (15 μm): is the broad and sloping absorption of the NH$_2$ wagging vibration.

with the carbonyl absorption, resulting in a broad band between 1695 and 1626 cm^{-1} (5.90–6.15 μm). The spectra of primary and secondary amides, taken as the neat liquid and in the solid state show extensive hydrogen bonding between the N–H hydrogen and the carbonyl group. In solution, the degree of hydrogen bonding is reduced, producing a corresponding shift of the carbonyl absorption to slightly higher frequencies. The carbonyl stretching vibrations of some amides, in solution and in the solid state, are shown in Table 2.4. These values are of a general nature, since factors such as structural features, electron-withdrawing and -donating effects and the nature of the solvent also influence the position of the absorptions.

Table 2.4 Some carbonyl absorptions of amides in the solid state and in solution

	Solid state (nujol mull)	Solution
Primary amides	Near 1650 cm^{-1} (6.60 μm)	1689 cm^{-1} (5.92 μm)
Secondary amides	Near 1640 cm^{-1} (6.10 μm)	1689–1700 cm^{-1} (5.92–5.88 μm)
Tertiary amides	1680–1630 cm^{-1} (5.95–6.13 μm)	1647–1615 cm^{-1} (6.07–6.20 μm)

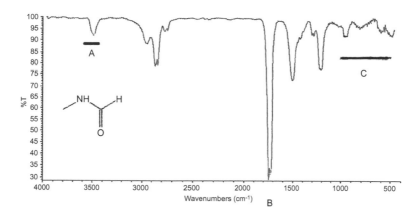

Figure 2.29 N-Methylformamide, liquid film. (A) 3333 cm^{-1} (3.0 μm): N–H stretching vibration. (B) Note the distinct separation of the C=O stretching and N–H bending vibrations, near 1724 cm^{-1} (5.8 μm) and 1515 cm^{-1} (6.6 μm) respectively, sometimes these two absorptions will overlap to give one broad absorption (Fig. 2.28). (C) Near 753 cm^{-1} (13.6 μm): the N–H wagging vibration.

The appearance and position of the N–H stretching vibration of amides depend, in part, upon structural features of the amide and its physical state. For instance, primary amides in dilute solutions, usually show symmetrical and asymmetrical H–N–H stretching vibrations near 3520 and 3400 cm^{-1} (2.84 and 2.94 μm) respectively. Secondary amides in dilute solutions show only one band which appears near 3333 cm^{-1} (3 μm). Naturally, this absorption is absent in the spectra of tertiary amides. In the solid state the N–H stretching vibrations of primary amides occur near 3350 and 3180 cm^{-1} (2.99 and 3.15 μm). For secondary amides, taken as the solid sample, the N–H absorptions usually appear as multiple bands near 3330 cm^{-1} (3.0 μm).

The N–H bending vibrations of primary amides occur around 1615 cm^{-1} (6.2 μm). As previously mentioned, this absorption often overlaps with the carbonyl stretching vibration resulting in a broad absorption in the 1613 cm^{-1} (6.2 μm) region (Fig. 2.28). Wagging vibrations of the N–H group are usually broad and sloping and centered near 690 cm^{-1} (14.5 μm), as shown in Figs. 2.28 and 2.29.

The C–N stretching vibration of amides (just as for amines) appears between 1175 and 1055 cm^{-1} (8.5–9.5 μm).

2.15 Esters

Esters are characterized by strong stretching vibrations of the C=O and C–O groups. The carbonyl stretching vibration generally occurs at slightly higher frequency than those of aldehydes and ketones, appearing between 1750 and 1753 cm^{-1} (5.71–5.76 μm). Some approximate carbonyl stretching vibrations of esters are shown below.

Some approximate carbonyl absorptions of esters

–C=C–CO–O–	1720 cm^{-1} (5.81 μm)	Conjugation results in very small shifts
AR–CO–O–	1720 cm^{-1} (5.81 μm)	
–CO–O–C=C–	1760 cm^{-1} (5.68 μm)	
AR–CO–O–Ar	1735 cm^{-1} (5.76 μm)	
α–Keto ester	1745 cm^{-1} (5.73 μm)	
β–Keto ester	1745 cm^{-1} (5.73 μm)	β–Keto groups have no affect on the ester carbonyl

Conjugation with the aromatic ring shifts the carbonyl absorption to lower frequency, where it appears between 1730 and 1715 cm^{-1} (5.76–5.83 μm). The C–O stretching vibration of esters usually appears as two broad and strong absorptions between 1300 and 1000 cm^{-1} (7.7–10 μm) as shown in the spectrum of ethyl acetate (Fig. 2.30).

2.16 Lactones

Carbonyl stretching vibrations of lactones (cyclic esters) and unconjugated aliphatic esters occur in the same general region (Fig. 2.31). For five-membered lactone rings, the C=O stretching vibration appears between 1885–1820 cm^{-1} (5.3–5.5 μm). As the size of the ring increases, the absorption is shifted toward lower frequency. When the C=O group of a five-membered ring lactone is conjugated, the carbonyl absorption appears between 1786 and 1754 cm^{-1} (5.6–5.7 μm). Carbonyl stretching vibrations of unconjugated six-membered ring lactones, cannot be distinguished from unconjugated esters (compare Figs. 2.30 and 2.31).

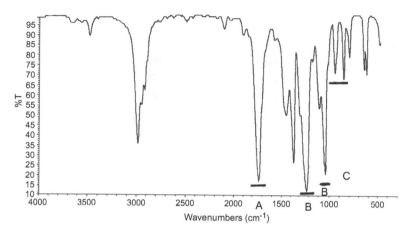

Figure 2.30   Ethyl acetate, neat. (A) 1736 cm^{-1} (5.76 µm): C=O stretching vibration. (B) 1250 and 1051 cm^{-1} (8.00 and 9.51 µm): C–O stretching vibration (two asymmetrical coupled vibrations). (C) 1000–7.14 cm^{-1} (10–14 µm): portions of the "fingerprint" region.

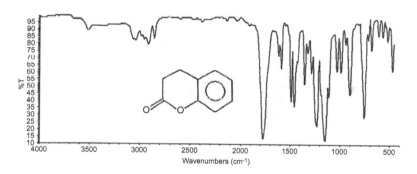

Figure 2.31   Dihydrocoumarin, neat.

2.17 Anhydrides

Anhydrides are characterized by two C=O stretching vibrations; (asymmetrical and symmetrical) located between 1886 and 1725 cm^{-1} (5.3–5.8 µm). For non-cyclic anhydrides, the absorption at higher frequency is the more intense of the two (compare Figs. 2.32, 2.33, and 2.34). Conversely, for cyclic anhydrides, the absorption at higher frequency is of lower intensity) (Table 2.5 and Figs. 2.32, 2.33,

Table 2.5 Carbonyl absorptions of some cyclic and open chain anhydrides

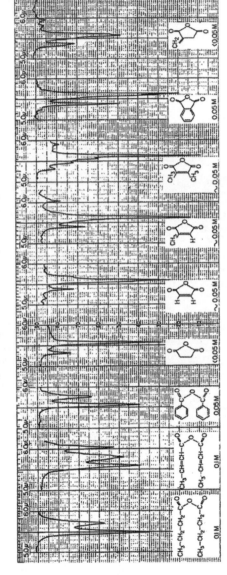

Source: Freeman, S. K. (editor), *Interpretive Spectroscopy*, Chapter 2, New York, Reinhold Publishing Company (1965).

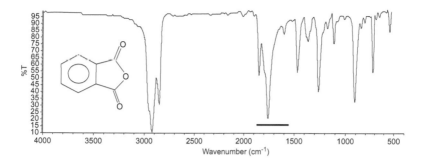

Figure 2.32   Phthalic anhydride, nujol mull.

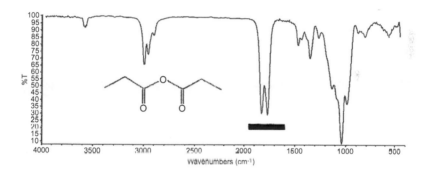

Figure 2.33   Propionic anhydride, liquid film.

and 2.34). *These differences make it possible to distinguish among cyclic and non-cyclic anhydrides.*

The carbonyl stretching vibrations of non-cyclic aliphatic anhydrides usually appear near 1750 cm^{-1} (5.50 μm), but when conjugated with a double bond, the absorption is shifted to near 1775 cm^{-1} (5.63 μm) and 1720 cm^{-1} (5.81 μm). As shown in Fig. 2.34, cyclic anhydrides containing five-membered rings, show the carbonyl stretching vibration near 1865 cm^{-1} (5.37 μm) and 1782 cm^{-1} (5.62 μm).

Aromatic anhydrides also show the characteristic carbonyl doublet (Fig. 2.35); however, in some cases it is not as pronounced as it is for aliphatic anhydrides. For aromatic anhydrides, the carbonyl absorptions appear between 1885 and 1725 cm^{-1} (5.3–5.8 μm), which is in the same region as the carbonyl absorption of aliphatic

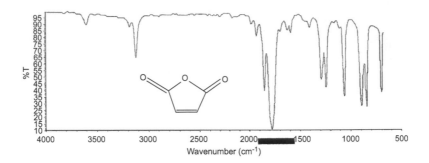

Figure 2.34   Maleic anhydride.

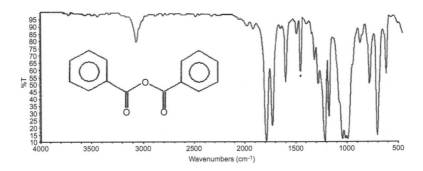

Figure 2.35   Benzoic anhydride, nujol mull.

anhydrides. The C–O stretching vibration for both aromatic and aliphatic non-cyclic anhydrides appears between 1250 and 1000 cm^{-1} (8–10 μm). For cyclic anhydrides, this absorption is shifted to approximately 910 cm^{-1} (11 μm). In most cases, the C–O stretching vibration is broad and strong.

2.18 Nitriles

The C≡N stretching vibration of nitriles appears between 2260 and 2240 cm^{-1} (4.42–4.46 μm), as a weak to strong absorption. Rarely is this absorption obscured by other bands in the region (Fig. 2.36). For aromatic nitriles, the C≡N stretching vibration is shifted to slightly lower frequency and appears between 2242 and 2222 cm^{-1}

Table 2.6   The stretching vibrations of multiple bonds

Nitriles	$-C{\equiv}N$	2260–2240 cm^{-1} (4.42–4.46 µm)
Alkynes	$-C{\equiv}C$	2235–2210 cm^{-1} (4.47–4.52 µm)
Alkenes (*cis* and vinyl)	$-C{=}C$	1660–1620 cm^{-1} (6.02–6.17 µm)
Alkenes (*trans*, tri and tetra substituted)	$-C{=}C$	1680–1660 cm^{-1} (5.95–6.17 µm)
Imine	$-N{=}C$	2175–2130 cm^{-1} (4.60–4.70 µm)

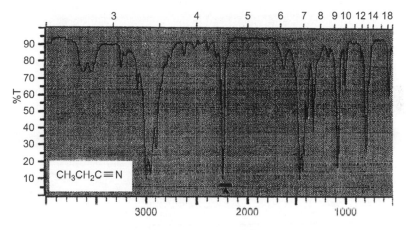

Figure 2.36   Propanenitrile, liquid film. (A) 2240 cm^{-1} (4.46 µm): the –C≡N stretching vibration. (B) As Table 2.6 indicates, the position of the –C≡N stretching vibration makes it possible to distinguish among, C≡N C=C, and N=C bonds.

(4.46–4.50 µm). The intensity of the C≡N stretching absorption is influenced by electronegative groups within its vicinity. In cases where the alpha carbon is substituted by an electronegative group, the intensity of the C≡N stretching absorption is either weak or completely absent. *Based upon the positions of multiple bond stretching vibrations, it is usually possible to make distinctions among nitriles, imines and alkenes (Table 2.6).* However, the C≡C and C≡N stretching vibrations occur in the same general region, making it difficult to distinguish between these two classes of compounds solely on the basis of multiple bond stretching vibrations.

2.19 Aromatic Hydrocarbons

The C–H stretching vibrations of aromatic hydrocarbons occur between 3100 and 3000 cm^{-1} (3.23–3.33 μm), while aliphatic C–H stretching vibrations appear between 2962 and 2853 cm^{-1} (3.38–3.51 μm). More prominent than the C–H aromatic stretching vibration is the medium to strong C–H in-plane bending vibration, located between 1300 and 1000 cm^{-1} (7.70–10.0 μm).

As shown in Figs. 2.37, 2.38, 2.39, and 2.40, aromatic C⋯C stretching vibrations appear between 1665 and 1430 cm^{-1} (6 and 7 μm). In this region, there are often two sets of multiple absorption bands of moderate intensity. One set is centered around 1600 cm^{-1} (6.25 μm) and the other between 1515 and 1430 cm^{-1} (6.6–7.0 μm). Sometimes the latter absorption is obscured by aliphatic C–H bending vibrations. *An absence of absorptions in the 1665–1430 cm^{-1} (6–7 μm) region may be taken as proof that the compound is non-aromatic.* Rocking vibrations of aromatic hydrogens occur between 1250 and 1000 cm^{-1} (8–10 μm). *An important group of absorptions common to substituted benzenes appears between 1000*

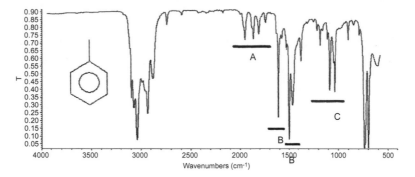

Figure 2.37 (A) 1923–1724 cm^{-1} (5.2–5.8 μm): overtone and combination bands, suggesting monosubstitution (see Fig. 2.41 and Table 2.7). (B) 1601 and 1500 cm^{-1} (6.25 and 6.67 μm): symmetrical ring stretch. (C) 1250–769 cm^{-1} (8–13 μm): "fingerprint" region. In-plane CH bending is found in this region but the absorptions are variable. (D) 728 cm^{-1} (13.7 μm): symmetric out-of-plane bending of the ring hydrogens suggesting a monosubstituted ring (see Table 2.7 and Fig. 2.41). (E) 728 and 693 cm^{-1} (13.7 and 14.4 μm): aromatic C–H out-of-plane bending vibrations, suggesting a monosubstituted ring (see Table 2.7 and Fig. 2.41).

Table 2.7 Aromatic C–H out-of-plane bending vibrations for some substituted benzene

Substitution patterns	Absorption (cm^{-1})
Benzene	671
Monosubstitution	770–730 & 710–690
1,2	770–735
1,3	810–750 & 710–690
1,4	833–810
1,2,3	780–760 & 745–705
1,2,4	825–805 & 885–870
1,3,5	865–810 & 730–765
1,2,3,4	810–800
1,2,3,5	850–840
1,2,4,5	870–855
Pentasubstitution	870

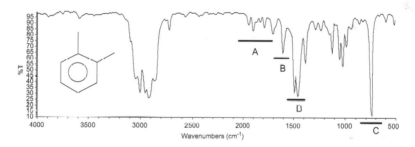

Figure 2.38 *o*-Xylene, liquid film. (A) 2000–1667 cm^{-1} (5.0–6.9 μm): overtone and combination band, suggesting *ortho*-disubstitution (see Fig. 2.41 and Table 2.7). (B) 1600–1490 cm^{-1} (6.2–6.7 μm): ring stretching vibrations which are characteristic bands of aromatic compounds. (C) 730 cm^{-1} (13.7 μm): symmetric out-of-plane bending vibrations of the aromatic ring, suggesting *ortho*-disubstitution (Table 2.7).

and 670 cm^{-1} (10–15 μm). These absorptions are the result of C–H out-of-plane bending modes and may vary in number and intensity depending upon ring substitution patterns. *If there are two absorption bands near 735 and 694 cm^{-1} (13.6 and 14.4 μm), the benzene ring is probably monosubstituted (Figs. 2.37). If there is no absorption near 694 cm^{-1} (14.4 μm), the sample compound cannot be a monosubstituted benzene. If only one absorption appears near 750 cm^{-1} (13.3 μm), there is a good chance that the compound*

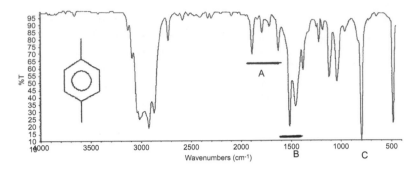

Figure 2.39 *p*-Xylene, liquid film. (A) 2000–1667 cm^{-1} (5–6 µm): overtone and combination bands, suggesting *p*– disubstitution (see Fig. 2.41 and Table 2.7). (B) 1515 cm^{-1} (6.6 µm): ring stretching vibrations. (C) 735 cm^{-1} (13.6 µm): symmetric out-of-plane bending vibrations of the ring hydrogens. The position of this band suggests p-disubstitution of the benzene ring (Table 2.7).

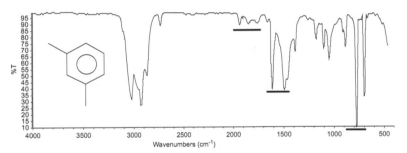

Figure 2.40 *m*-Xylene, liquid film.

is orthodisubstituted (Figs. 2.38). Table 2.7 lists the expected out-of-plane bending modes for certain ring substitution patterns. This data is not 100% reliable and should be used with this understanding.

If the infrared spectrum is of a substituted benzene, taken as the neat liquid, if may also be possible to verify the substitution pattern on the ring by matching the overtone and combination bands. As shown in Figs. 2.37, 2.38, 2.39, 2.40, and 2.41, these very weak bands are located between 2000 and 1667 cm^{-1} (5.0–6.0 µm). The patterns of these overtone and combination bands will compliment the C–H out-of-plane bending in arriving at the substitution pattern of the ring.

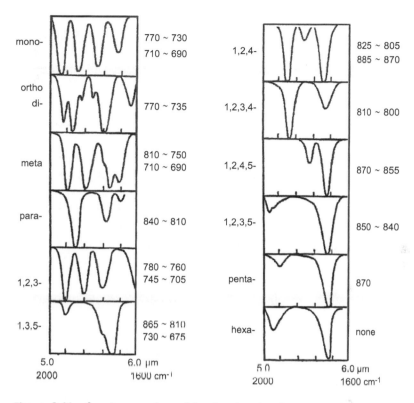

Figure 2.41 Overtone and combination bands, showing aromatic ring substitution pattern.

How to Determine If a Compound Is Aromatic

In order to determine if the infrared spectrum is that of an aromatic compound, the analyst should consider the following significant infrared absorptions.

A. **The C–H Aromatic Ring Stretching Vibrations:** These bands are of medium to weak intensities and are always found above 3000 cm^{-1} (below 3.3 μm). Specifically, these absorptions occur between 3125 and 3030 cm^{-1} (3.2–3.3 μm). Again, when there is an absorption above 3000 cm^{-1}, the compound has some form of C–H unsaturation (aromatic ring, alkene or alkyne).

B. **Overtone and Combination Bands:** These weak overtone and combination bands are located between 2000 and 1700 cm^{-1} (5–5.9 μm). The bands are only useful for the characterizing of liquids samples. The group of bands is indicative of substitution patterns on benzene rings as shown above and below.

C. **Symmetrical In-Plane Aromatic Ring Stretching Vibrations:** This absorption generally occurs between 1600 and 1580 cm^{-1} (6.25–6.33 μm). The absorption is absent when the ring is symmetrically substituted and its intensity varies with the substituent.

D. **More In-Plane Aromatic Ring Stretching Vibration** with variable intensity occurs between 1510 and 1490 cm^{-1} (6.62–6.71 μm).

E. **More In-Plane Aromatic Ring Stretching Vibration** with variable intensity occurs between 1440 and 1460 cm^{-1} (6.94–6.85 μm). A kind of "sideways" ring stretching vibration. This band often overlaps with the CH$_3$ bending vibration and is therefore of little diagnostic value.

F. **Symmetrical Out-Of-Plane Bending of the Ring Hydrogens:** These absorptions are found between 750 and 710 cm^{-1} (13.33–14.08 μm). There appears to be some correlation with the number of adjacent ring hydrogens and the position of the outof-plane bending of the ring hydrogens. As shown below, the greater the number of adjacent ring hydrogens on the aromatic ring, the lower the frequency of the out-of-plane bending vibration. This information should be considered very carefully since minor structural features may alter the position of the aromatic C–H

Compound	Adjacent ring hydrogens	Position of absorption
Benzene	6	671 cm^{-1} (14.90 μm)
Toluene	5	732 cm^{-1} (13.66 μm.)
Ethyl benzene	5	745 cm^{-1} (13.42 μm)
Chlorobenzene	5	740 cm^{-1} (8–10 μm)
o-Xylene	4	740 cm^{-1} (13.51 μm)
m-Xylene	3	775 cm^{-1} (12.90 μm)
p-Xylene	2	790 cm^{-1} (12.66 μm)
1,3,5-Triethyl benzene	1	865 cm^{-1} (11.56 μm)

aromatic outof-plane bending vibration. For instance, o-xylene and chlorobenzene show this absorption at the same position.

G. **Out-of-Plane Ring Bending:** This absorption is located between 710 and 670 cm^{-1} (14.08–14.92 μm). The absorption appears only for mono, meta and 1,3,5 trisubstituted benzenes.

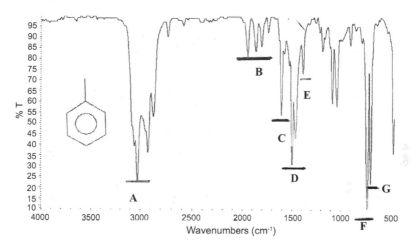

Summary of Ring Vibrations for Toluene

3100–3000 cm^{-1} (3.22–3.33 μm)	Unsaturated aromatic C–H stretching
2000–1700 cm^{-1} (5.00–5.88 μm)	Overtone pattern (for Mono substitution of the ring)
1600 cm^{-1} (6.25 μm)	Symmetrical in-plane ring stretching
1500 cm^{-1} (6.67 μm)	More in-plane ring stretching
1450 cm^{-1} (6.89 μm)	More in-plane ring stretching
730 cm^{-1} (13.70 μm)	Out-of-plane C–H bending
690 cm^{-1} (14.49 μm)	Out-of-plane ring bending

MISCELLANEOUS COMPOUNDS

2.20 Salts of Carboxylic Acid

Salts of carboxylic acids show a shift in the carbonyl absorption. In the free acid the carbonyl absorption appears near 1728 cm^{-1} (5.79 μm) and in the acid salt it appears between 1625 and 1540 cm^{-1}

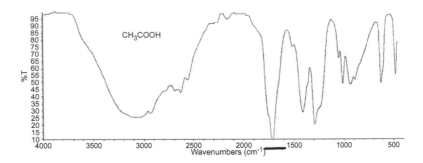

Figure 2.42 Acetic acid, liquid film.

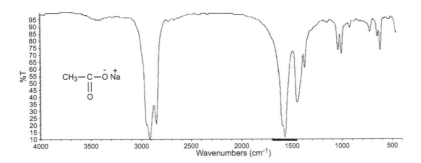

Figure 2.43 Sodium acetate, nujol mull.

(6.2–6.5 μm). As an example, compare the carbonyl absorption of acetic acid (Fig. 2.42) and sodium acetate (Fig. 2.43).

2.21 The Effects of Chelation on the Carbonyl Absorption

The effects of chelation on the carbonyl absorption may be illustrated by comparing the spectra of acetylacetone (Fig. 2.44) and a few of its chelates (Figs. 2.45, 2.46, and 2.47).

First of all, acetylacetone (Fig. 2.44) exists as an equilibrium mixture of enol and keto tautomers. The carbonyl group of the keto tautomer appears as a doublet between 1754 and 1695 cm^{-1}

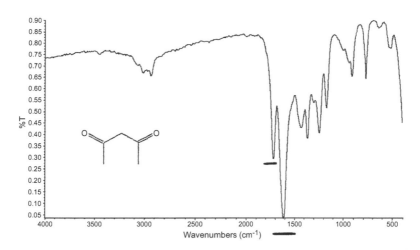

Figure 2.44 Acetylacetone (2,4-pentanedione), liquid film. (A) 1754 and 1695 cm^{-1} (5.7 and 5.9 µm): The carbonyl stretching vibrations of the keto form of the acetylacetone tautomer. (B) 1587 cm^{-1} (6.3 µm): the carbonyl absorption of the enol form of the acetylacetone tautomer.

(5.7 and 5.9 µm) and for the enol form, the carbonyl group appears as a single band near 1587 cm^{-1} (6.3 µm).

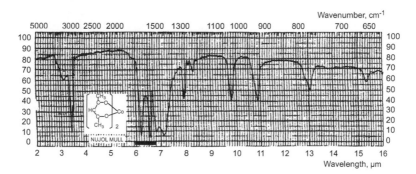

Figure 2.45 Calcium acetylacetonate, nujol mull. (A) 1615 and 1515 cm^{-1} (6.2 and 6.6 µm): the chelated carbonyl stretching vibration. (B) Compare the chelated carbonyl stretching vibration in Figs. 2.44, 2.45, and 2.46 with the carbonyl absorption of acetylacetone (Fig. 2.43). *Source*: The Aldrich Library of Infrared Spectra © Aldrich Chemical Company, Inc.

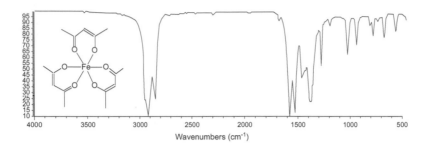

Figure 2.46 Ferric acetylacetonate, nujol mull.

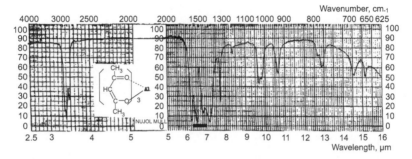

Figure 2.47 Aluminum acetylacetonate, nujol mull. *Source*: The Aldrich Library of Infrared Spectra © Aldrich Chemical Company, Inc.

$$CH_3-\overset{\overset{\displaystyle O}{\|}}{C}-CH_2-\overset{\overset{\displaystyle O}{\|}}{C}-CH_3 \rightleftharpoons CH_3-C=CH-\overset{\overset{\displaystyle O}{\|}}{C}-CH_3$$

Keto Form   Enol Form

In the chelated form of acetylacetone, the carbonyl absorption appears as a doublet and is shifted between 1640 and 1492 cm^{-1} (6.1–6.7 μm) as shown in Figs. 2.45, 2.46, and 2.47. In some acetylacetonates (tin and vanadium), the two absorptions coalesce to give a single absorption near 1540 cm^{-1} (6.5 μm).

2.22 Phosphines, Phosphites, Phosphonates and Phosphates

As shown below, the compounds of phosphorus include the phosphines, phosphites, phosphonates and phosphates.

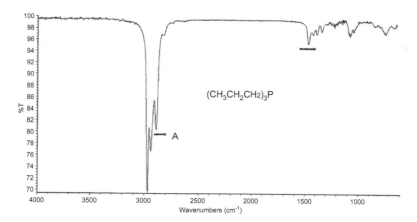

In phosphines, the P–C stretching vibration is not a good diagnostic absorption. In addition, attached phosphorus has very minimum effects on CH_2 bending vibrations, thus CH_2 bending vibrations of P–CH_2 groups occur in the same general region as most other CH_2 bending vibrations. As shown in Fig. 2.48, the CH_2 stretching of P–CH_2 groups often appears as a weak shoulder near 2815 cm^{-1} (3.55 μm). Thus, if not obscured, this absorption has some diagnostic value.

The P–H stretching vibration occurs near 2440 cm^{-1} (4.1 μm) as a characteristic band, free from interference by other bands in the

Figure 2.48 Tri-n-propylphosphine. (A) 2815 cm^{-1} (3.55 μm): this weak shoulder is attributed to the CH_2 stretch of the P–CH_2 group. (B) As noted, the P–C absorption is not a characteristic absorption of phosphines and thus has little diagnostic value. Also, the CH_2 deformation of the P–CH_2 group usually overlaps with the C–H bending vibrations near 1470 cm^{-1} (6.8 μm).

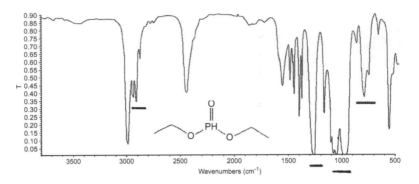

Figure 2.49 Diethylphosphite, neat. (A) 2440 cm^{-1} (4.1 μm): the P–H stretching vibration. (B) 1250 cm^{-1} (8.0 μm): the P=O stretching vibration. (C) The two strong absorptions between 1110 and 870 cm^{-1} (9–11.5 μm) can be attributed to the P–O–C group. (D) Near 741 cm^{-1} (13.5 μm): this medium intensity band occurs between 835 and 715 cm^{-1} (12–14 μm) and is associated with the P–O–C group.

region (Fig. 2.49). Also shown in Fig. 2.49 is the strong stretching absorption of the P=O group near 1250 cm^{-1} (8.0 μm). If a halogen is attached to the phosphorus as in Cl–P=O, the P=O band may be shifted to slightly lower wavenumbers.

The P–O–C group often shows two strong absorptions between 1110 and 1000 cm^{-1} (9–10 μm) and 1000–870 cm^{-1} (10.0–11.5 μm) as shown in Figs. 2.49 and 2.50. In some cases, the two absorptions may coalesce forming only one absorption which appears between 1110 and 870 cm^{-1} (9–11.5 μm).

The O–H stretching vibration of the P–O–H group appears as a strong absorption over a wide range, generally occurring between 3335 and 1540 cm^{-1} (3.0–6.5 μm). Within this range there are usually three maximums (Fig. 2.50), located around 2630, 2220 and 1665 cm^{-1} (3.8, 4.5 and 6.0 μm)

2.23 Thioalcohols and Thiophenols

The S–H stretching vibrations of thioalcohols (mercaptans) and thiophenols, when taken neat or in solution, appear as a weak absorption near 2565 cm^{-1} (3.9 μm). Contrast this absorption to

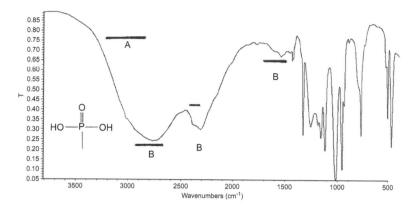

Figure 2.50 Methylphosphonic acid. (A) 3335–1540 cm^{-1} (3.0–6.5 μm): this broad and wide absorption is characteristic of the O–H stretching vibrations of the P–O–H group. (B) *Note*: the three maximums, located around 2630, 2220 and 1665 cm^{-1} (3.8, 4.5 and 6.0 μm) which are also characteristic of the P–O–H group.

the normal stretching vibration of the O–H group, which occurs near 3500–3200 cm^{-1} (3.13–2.75 μm) as a strong and broad absorption. Examples of the S–H stretching vibration are shown in Figs. 2.51, 2.52, and 2.53. Although weak, this characteristic absorption is often a good indicator for both thioalcohols and thiophenols. However, as the carbon chain of the thioalcohol increases beyond nine carbons,

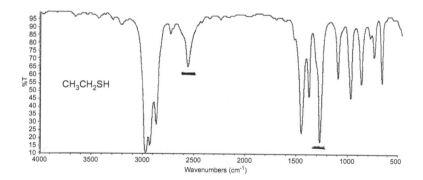

Figure 2.51 Ethanethiol (ethyl mercaptan). (A) 2565 cm^{-1} (3.9 μm): S–H stretching vibration. (B) 1250 cm^{-1} (8 μm): CH$_2$ wagging vibrations. This is the only characteristic absorption that can be related to divalent sulfur.

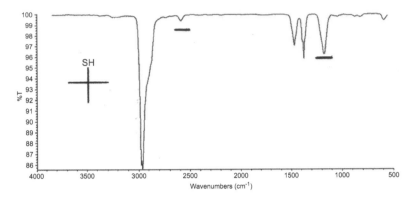

Figure 2.52 2-Methyl-2-propanethiol (t-butylmercaptan). (A) 2565 cm^{-1} (3.9 μm): S–H stretching vibration. *Note*: this absorption appears in the same general position for thiophenols (see Fig. 2.53). (B) 1250 cm^{-1} (8 μm): CH$_2$ wagging vibration.

the S–H stretching vibration may become unrecognized because of overtone bands in the same general region.

If sulfur is bonded to a CH$_2$ group as in S–CH$_2$, there is usually a characteristic absorption near 1250 cm^{-1} (8 μm) which is attributed to the CH$_2$ wagging vibrations (Figs. 2.51 and 2.52). *This medium absorption is one of the few that is characteristic of divalent sulfur.*

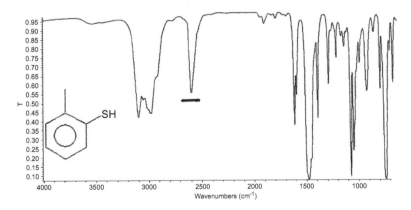

Figure 2.53 o-Thiocresol, neat. (A) 2565 cm^{-1} (3.9 μm): S–H stretching vibration. (B) The S–H stretching vibration is about the only good diagnostic absorption for thiophenols.

The C–S stretching vibration appears between 715 and 625 cm^{-1} (14–16 μm) as a weak absorption. It is often obscured by other absorptions in the area and is of little diagnostic value.

2.24 Silanes

The Si–H stretching vibration appears near 2130 cm^{-1} (4.7 μm) as a strong vibration (Fig. 2.54). Bending (deformation) vibrations of the Si–H groups appear between 950 and 800 cm^{-1} (10.5–12.5 μm). Si–O–C (aliphatic) groups appear as a strong vibration between 1110 and 1000 cm^{-1} (9–9.5 μm).

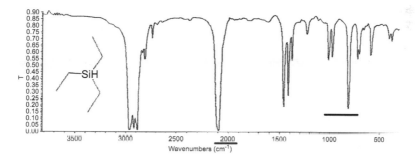

Figure 2.54 Triethylsilane. (A) 2130 cm^{-1} (4.7 μm): Si–H stretching vibration. (B) 950–800 cm^{-1} (10.5–12.5 μm): Si–H deformations.

For aromatic silanes, the Si–C stretching vibrations appear as two bands near 1430 and 1110 cm^{-1} (7 and 9 μm). These absorptions are shown in the spectrum of chlorodiphenylmethylsilane (Fig. 2.55).

2.25 Boranes

The structure of boranes salts may be illustrated as following:

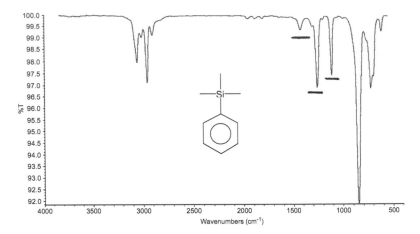

Figure 2.55 Trimethylphenylsilane. (A) 1430 and 1110 cm^{-1} (7 and 9 μm): Si–C (aromatic) stretching vibration. (B) 1250 cm^{-1} (8 μm): aliphatic Si–C stretching vibration of the SI–CH$_3$ group. This absorption is accompanied by the bands between 910 and 715 cm^{-1} (11–14 μm).

In these compounds, the B–H stretching vibration usually appears between 2500 and 2220 cm^{-1} (4–4.5 μm). As an example, consider the spectrum of borane trimethylamine complex in Fig. 2.56.

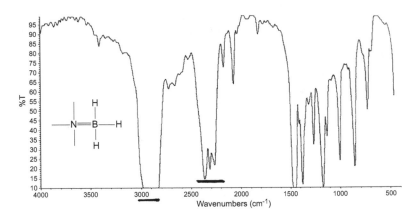

Figure 2.56 Borane trimethylamine complex. (A) 2380 cm^{-1} (4.2 μm): the B–H stretching vibration. (B) Note that the B–H stretching vibration is far removed from C–H stretching near 3030 cm^{-1} (3.3 μm).

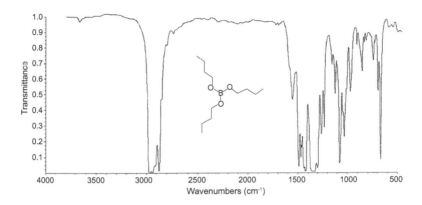

Figure 2.57 Tri-n-butylborate.

The B–O and B–N groups appears between 1430 and 1355 cm^{-1} (7.0–7.5 μm) as a strong absorption (Fig. 2.57). *When boron is attached directly to the aromatic ring, there is usually a medium to strong absorption near 1430 cm^{-1} (7 μm), which is due to the B–C (aromatic) group.* A similar band near 1430 cm^{-1} (7 μm) is also observed when P, Si, As, Sb, Sn and Pb are directly attached to the aromatic ring.

2.26 Sulfones, Sulfates, Sulfonic Acids (and Their Salts), Sulfites and Sulfoxides

The structures of some common sulfur containing compounds are shown below:

2.26.1 Sulfones

As shown by the spectrum of diphenylsulfone (Fig. 2.58), the sulfone group (O=S=O) appears as two bands. The asymmetrical stretching vibration appears as a strong absorption near 1300 cm^{-1} (7.7 μm), while the symmetrical stretch appears near 1135 cm^{-1} (8.8 μm), also as a strong band.

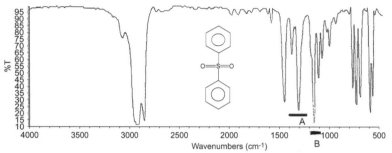

Figure 2.58   Phenylsulfone. (A) 1299 cm^{-1} (7.7 μm): asymmetrical stretching vibration of the O=S=O group. (B) 1135 cm^{-1} (8.8 μm): symmetrical stretching vibration of the O=S=O group.

2.26.2 Sulfates

Sulfates show two absorptions near 1390 and 1205 cm^{-1} (7.2–8.3 μm) due to the O=S=O stretching vibrations (Fig. 2.59). The S–O–C absorptions occur between 1055 and 770 cm^{-1} (9.5 and 13 μm).

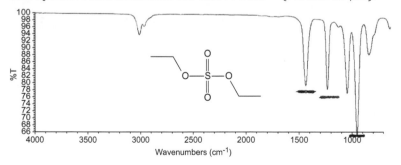

Figure 2.59   Diethyl sulfate, neat. (A) 1390 and 1205 cm^{-1} (7.2–8.3 μm): the O=S=O stretching vibrations of sulfates. (B) The group of bands between 1055 and 770 cm^{-1} (9.5 and 13 μm) can be attributed to the S–O–C group.

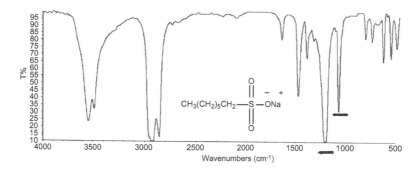

Figure 2.60 1-Heptanesulfonic acid, sodium salt.

2.26.3 Sulfonic Acids (and Their Salts)

The spectrum of alkyl sulfonic acid salts is typified by 1-heptanesulfonic acid, sodium salt (Fig. 2.60). As noted, the O=S=O group appear as a strong band near 1175 cm^{-1} (8.5 μm) and as a medium band near 1055 cm^{-1} (9.5 μm). These two absorptions are also present in the hydrated form of the free alkyl sulfonic acid.

2.26.4 Sulfites

The S=O stretching vibrations of sulfites appear near 1205 cm^{-1} (8.3 μm) as shown for diethyl sulfite (Fig. 2.58). The S–O–C group of sulfites appear as two strong bands near 1000 and 910 cm^{-1} (10 and

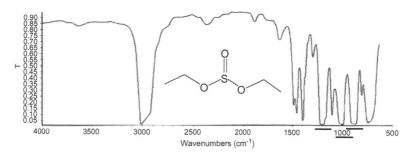

Figure 2.61 Diethyl sulfite, neat. (A) 1205 cm^{-1} (8.3 μm): C=O stretching vibration. (B) 1020 cm^{-1} (9.8 μm) and 885 cm^{-1} (11.3 μm): these two absorptions are attributed to the S–O–C group.

11 μm). Another pair of bands common to sulfites occur between 770 and 665 cm^{-1} (13–15 μm).

2.26.5 Sulfoxides

Aryl and alkyl sulfoxides show strong and broad absorption between 1070 and 1030 cm^{-1} (9.5–10 μm) as shown in Figs. 2.62 and 2.63.

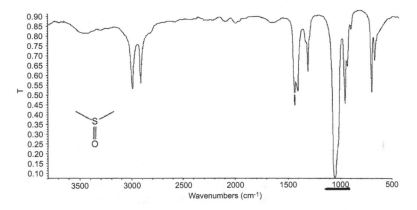

Figure 2.62   Dimethyl sulfoxide, neat.

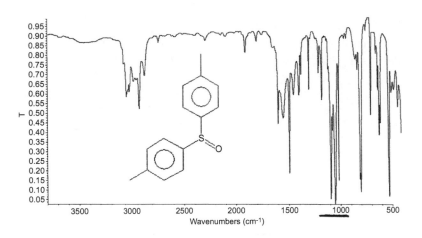

Figure 2.63   p-Tolyl sulfoxide, nujol mull.

Chapter 3

Techniques, Innovations, and Applications in Infrared Spectroscopy

3.1 Advances in Infrared Spectroscopy

In recent years, infrared spectroscopy has experienced a vast improvement in many areas. Among these are resolution, sensitivity, speed, the development of innovative IR accessories, analysis software, micro sampling and new analytical techniques. These improvements may be attributed to two major factors: (a) the merger of the Michelson Interferometer with the mathematics–computer technique of fast Fourier transform (FFT) and (b) advances in computer technology, allowing infrared spectrometers to be interfaced with microcomputers for data acquisition, analysis, processing and spectral searching. FTIR has become a valuable analytical procedure in many areas of fundamental research and quality control. The advances in infrared spectroscopy have made it possible to obtain routine infrared spectra on almost any material, often with little or no sample preparation.

Infrared Spectroscopy
James M. Thompson
Copyright © 2018 Pan Stanford Publishing Pte. Ltd.
ISBN 978-981-4774-78-9 (Hardcover), 978-1-351-20603-7 (eBook)
www.panstanford.com

3.2 Fast Fourier Transform (FFT)

Fast Fourier transform (FFT) is a mathematical–computer procedure that can process and convert information from an interferogram into a typical infrared spectrum. The FFT procedure makes use of the Cooley–Tukey algorithm to perform the transformation. Unlike dispersive instruments, the FTIR spectrometers allow the entire frequency range of a spectrum to be scanned in microseconds. Therefore, Fourier Transform Infrared (FTIR) may be used to obtain information on dynamically changing systems such as evolved gases resulting from thermal decompositions and fractions eluting from the column of a gas chromatograph. Advances in FTIR have also made it possible to couple FTIR spectrometers with other instruments such as thermoanalyzers, gas chromatographs and mass spectrometers. The information obtained from these coupled systems is usually greater than the sum of information obtained from the systems operating separately.

3.3 The Michelson Interferometer

In 1887 Albert Michelson, an American scientist born in Germany perfected the interferometer. Among other things, he used the optical device to study the properties of light. However, the application of the interferometer to infrared spectroscopy had to await improvements in computer technology and the development of FFT, both of which took place some years later. Shown schematically in Fig. 3.1, the interferometer consists of a beamsplitter that splits the source radiation about equally into transmission and reflected radiation. Both the transmitted and reflected beams are directed to mirrors and reflected back to the beamsplitter where they recombine. Recombination at the beamsplitter occurs either with destructive or constructive interference, depending upon the differences in path lengths of the recombining beams. If the path lengths differ by some integer number of wavelength, the beams will recombine with constructive interference. If the difference is some non-integer value, recombination will result in destructive

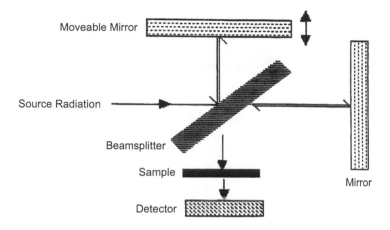

Figure 3.1   Schematic diagram of the Michelson interferometer.

interference. After recombining at the beamsplitter, the beam passes through the sample and onto the detector.

The ability to move one of the reflecting mirrors is crucial to the resulting interferometer pattern. If the source radiation is of one frequency, the resulting interferogram produces a sine wave, the spacing of which is dependent upon the change in path length and equal to the wavelength. For polychromatic radiation, the interferogram is more complex and has to be digitized and transformed using the FFT technique. When the FFT technique is applied, the relationship between the interferogram and the related set of sine and cosine frequencies is revealed. After subtracting the background spectrum, a typical infrared spectrum of the sample is produced in which the energy of infrared radiation is plotted as a function of frequency. The entire FFT process takes place in microseconds. Because of the nature and capability of the interferometer, most modern infrared instruments are of the FTIR type. These instruments have a decided advantage over the classical slit and grating based dispersive systems. Among the advantages are the following:

- FTIR instruments are faster since they are capable of measuring all frequencies simultaneously. The dispersive instrument measures frequencies successively. This advantage enables FTIR instruments to monitor dynamically changing systems.

- A helium–neon laser is used to calibrate FTIR spectra. The accuracy and stability of this calibrating procedure give rise to highly accurate absorption band measurements.
- The energy throughput is higher in the FTIR instrument, thus, it can achieve signal to noise ratios in shorter time.
- For the FTIR, resolutions at all wavelengths are equal. In the dispersive instrument, resolution varies because it is slit controlled.
- The nature of the interferometer is such that there is no stray light as it is with the dispersive instrument.
- The absence of grating and filters result in a continuous spectrum, whereas in the dispersive instruments there are discontinuities resulting from grating and filter changes.

3.4 The FTIR Microscope and Microsampling Techniques

Probable the most significant advance in FTIR microsampling techniques was the introduction, in 1983, of the FTIR microscope as a commercial accessory. Since its introduction, the technique has enjoyed considerable popularity for characterizing small samples including those in the picogram range. This popularity is such that FTIR microscopes are now made available by most major FTIR manufacturers.

In a typical FTIR microscope, the infrared beam travels through the sample which is located on the stage of the microscope (Fig. 3.2). The beam is then collected and focused on a mercury cadmium telluride (MCT) detector that is cooled by liquid nitrogen. The MCT detector is ideally suited for small samples with limited transmission properties. It operates on a photoconductive principle and is noted for its fast response, low noise level and exceptional sensitivity. The MCT and DTGS (deuterated triglycine sulfate) detectors are the two commonly used FTIR detectors. The DTGS detector is less sensitive than the MCT detector and is used at room temperature.

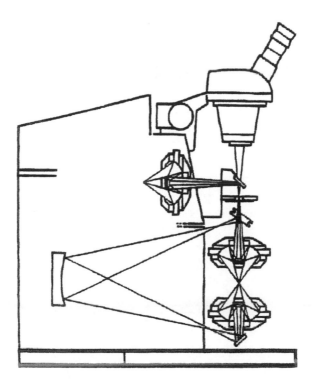

Figure 3.2 A schematic representation of an FTIR microscope. *Source*: Perkin–Elmer Corporation, Norwalk, CT.

The microsample is viewed through the low power objective of the microscope before the spectrum is obtained. The cassegrain is then switched on to increase magnification. Once the sample is in the viewing area, it is masked using a variable aperture. The spectrum is then taken on that portion of the sample within the mask. After obtaining the spectrum of the microsample (or any sample), it may be searched against a digitized reference library for structural matches. Among the specialized reference libraries commercially available include spectra of monomers, polymers, surfactants, adhesives, sealants, abused drugs, forensic materials, pharmaceuticals, common solvents, food additives, flavors and fragrances, automobile paint chips, steroids, inorganic and organic compounds. In addition, large libraries-containing over 200,000 digitized spectra are commercially available.

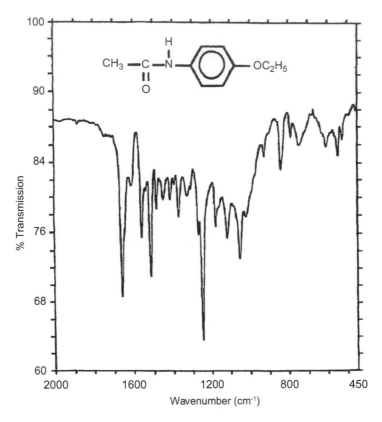

Figure 3.3 The infrared spectrum of 1 μg of phenacetin pressed into a 1.5 mm KBr pellet. Obtained on the Perkin–Elmer 1700 FTIR spectrophotometer using a DTGS detector. *Source*: Perkin–Elmer Corporation, Norwalk, CT.

Infrared spectra of microsamples may also be obtained using a beam condenser. As an example, consider the transmission spectrum of a 1 μg sample of phenacetin (Fig. 3.3) which was obtained using a commercially available beam condenser. As noted, the spectrum shows the presence of the amide carbonyl group near 1650 cm^{-1} (6.06 μ) and a series of absorptions characteristic of the aromatic ring. Beam condensers have been used to analyze microsamples such as spots on polymer films, single fibers and small paint chips. The accessory is relative inexpensive when compared to an FTIR microscope, but it is less versatile.

3.5 Reflective Spectra

With the proper accessories, FTIR spectrometers can produce both transmission and reflective spectra. The latter technique is useful in the study of non-transmitting materials and materials that scatter infrared radiation, such as pharmaceuticals, food products, detergents, coals, clays, paper products, painted surfaces, polymer foams and many other non-transmitting materials. Reflective measurements require little or no sample preparation and the technique has become a common FTIR procedure. The extent in which infrared radiation is reflected off the surface of a sample depends upon the following factors:

- Angle of the incident beam
- Thickness of the coating or film
- Index of refraction of the sample
- Roughness of the surface
- Absorption properties of the sample

Depending upon the sample characteristics, reflective measurements may be classified as follows:

- **Specular or External Reflectance**: Reflections from shiny or mirror-like surfaces.
- **Grazing Incidence Reflectance**: Similar to specular reflectance, but the angle of incidence is larger (80°)
- **Reflective-Absorptions or Double Pass Transmission Spectra**: Usually involves thin transparent films on reflecting surfaces, such as organic materials, and adhesives on metallic surfaces or oil on a smooth reflective surface.
- **Diffuse Reflectance**: Involves samples with irregular surfaces or aggregates of fine powdered materials.

Unlike transmission spectra, some reflective spectra (not all) produce distorted spectral bands, which may be attributed in part to the dispersion of the refractive index of the sample. However, the spectra can usually be software corrected (transformed) to produce "normal" transmission-like spectra.

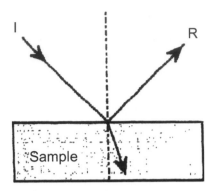

Figure 3.4 A schematic diagram showing the specular reflectance process. In pure specular reflectance, the angle of incidence (I) is Equal to the angle of reflectance (R).

3.6 Specular or External Reflectance

When infrared radiation is reflected off the smooth plane surface of a polished mirror-like material, specular reflectance spectra results (Fig. 3.4). There is little or no transmission of radiation involved in pure specular reflectance and the angle of reflection (R) is equal to the angle of incidence (I). As stated, the amount of infrared radiation reflected depends upon the absorption properties of the sample, the roughness of the surface, the angle of incidence and the index of refraction of the sample. Surfaces that are smooth and highly reflective are ideal candidates for specular reflectance measurements. For micron thick samples on reflective surfaces, reflective-absorption measurements are recommended. For submicron films on reflective surfaces, grazing incidence reflectance measurements give the best results.

Sample materials suitable for specular reflectance include surface-treated metals, resin and polymer coatings, paints on soda cans and semiconductors, to name a few. Specular reflective measurements often produce derivative-like spectra (Fig. 3.5a), that must be software corrected or transformed. This is accomplished by applying the mathematics–computer transformation called the Kramers–Kronig dispersion formula. Once transformed, a specular reflectance spectrum is converted to transmission type spectra

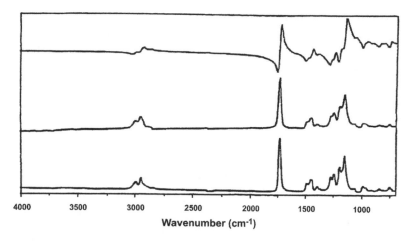

Wavenumber (cm⁻¹)

Figure 3.5 The spectrum of polymethyl methacrylate (PMMA) obtained by specular reflection (a) after the Kramers–Kronig transformation (b) and the transmission spectrum (c). *Source*: Reffner, J.A, and W.T. Wihlborg, *Am. Lab.*, April, 1990.

(compare spectra (a) and (b) in Fig. 3.5). The Kramers–Kronig transformation works best on pure specular reflectance spectra. If combinations of different reflective type spectra are present, interpretation becomes more difficult. The Kramers–Kronig transformation once required calculations afforded only by mainframe computers. With the arrival of powerful microcomputers, this requirement has changed. Today the procedure is routine and the software required to carry out the transformations is included in most standard commercial packages. The technique is nondestructive and requires no sample preparation; however, it does require a special accessory.

3.7 Grazing Incidence Reflectance

The thickness of the surface coating dictates the selection of the angle of incidence of the radiation beam. For samples with very thin surface coatings (in the nanometer range), better results are obtained if the angle of incidence is around 80°. Reflectance measurements at this angle are called grazing incidence reflectance.

If the surface coating is greater than 1 micron, best results are obtained if the angle of incidence is 30°. Both grazing angle and specular reflectance accessories are commercially available from a number of vendors.

3.8 Reflective–Absorption Spectra

The reflective–absorption technique (also called double pass transmission spectra) is suited for thin transparent materials on the surface of smooth flat reflecting surfaces (a transparent material on a mirror). In this procedure, the incident beam passes through the sample and is then reflected off the metal surface and back out of the sample (Fig. 3.6). The process results in transmission spectra and may be applied to transparent samples with a thickness between 0.5 and 20 microns.

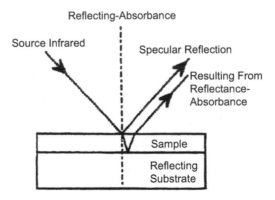

Figure 3.6   A Schematic diagram of the reflection–absorption process.

3.9 Diffuse-Reflectance Spectra

Diffuse-reflectance spectra result when the sample bas an irregular surface such as a fine powdered like material. Here some of the infrared radiation penetrates the sample to varying depths and is then reflected out of the sample at various angles, giving rise to

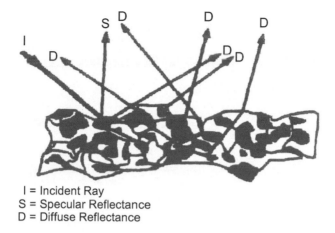

I = Incident Ray
S = Specular Reflectance
D = Diffuse Reflectance

Figure 3.7 Diffuse reflectance.

the diffuse-reflectance component (Fig. 3.7). Some of the infrared radiation is reflected directly off the sample (there is no absorption by the sample) as the specular reflection component (Fig. 3.7). The optics of the diffuse-reflectance accessory are designed to collect the scattered (diffuse) radiation and direct it to the infrared detector, while at the same time minimizing the specular reflectance component of the radiation.

To obtain a diffuse reflectance spectrum, the powered sample material is converted to a very fine powder and then mixed with a non-absorbing material such as KBr. By diluting the sample with a non-absorbing material, the percentage of the infrared radiation that is diffusely reflected is increased and a useful transmission-like spectrum such as the one shown in Fig. 3.8 is obtained. High concentrations of the sample material increase specular reflections and decrease diffuse reflections, making the spectrum difficult to interpret.

Diffusion reflectance spectra may be obtained from intractable and hard samples. This is achieved by abrading the sample material with a disk of silicon carbide or diamond paper (for extremely hard samples). The technique causes a small amount of the sample material to adhere to the carbide or diamond paper and a diffuse reflectance spectrum of the adhering material obtained.

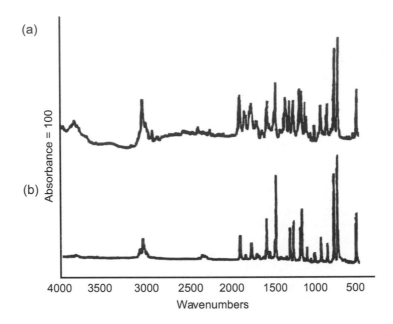

Figure 3.8 Comparison of the diffuse reflectance (a) and microtransmission (b) spectra of perylene ($C_{20}H_{12}$). *Source*: Krishnan, K., and S. L. Hill, Notes No. 73, Bio–Rad, Digilab Division, April, 1990.

3.10 Attenuated Total Reflectance (ATR)

Infrared studies of aqueous solutions were once prohibited due to the strong infrared absorption of water and the absence of suitable water insoluble infrared cells. With the development of innovative infrared cells such as the Attenuated Total Reflectance Cell (ATR Cell) and the Horizontal Attenuated Total Reflectance Cell (HATR Cell), the prohibition has disappeared. These specialized cells (Fig. 3.9) are fabricated with water insoluble crystalline materials with high refractive indices, such as zinc selenide (ZnSe) and germanium (Ge). Using the ATR technique, it is now possible to obtain spectra of a variety of aqueous and highly absorbing materials as well as pastes, gels, powders, films, fibers and rigid plastics. Next to the FTIR microscope, ATR is probably the next most versatile technique. The technique is fast and requires little or no

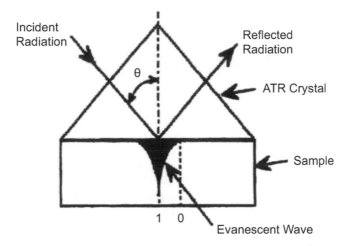

Figure 3.9 A schematic drawing of an attenuated internal reflectance (ATR) cell.

sample preparation and may be applied to a wide variety of sample types.

3.11 Theory of the ATR Cell

When infrared radiation enters certain crystalline materials with high refractive indices (such as ZnSe and Ge), the radiation is totally internally reflected. In addition, the internal reflectance creates a wave called an evanescent wave that extended slightly beyond the surface of the ATR crystal (Fig. 3.10). If a sample material is placed in contact with the ATR crystal, the evanescent wave penetrates the sample to a very small depth, producing an infrared spectrum at the surface of the sample material. The fact that the evanescent wave decays exponentially with distance from the surface of the ATR crystal is crucial to the ATR technique. This characteristic makes the ATR technique insensitive to the thickness of the sample. Thus, spectra of thick and highly absorbing materials may be obtained using the procedure. With the development of horizontal ATR crystals multiple internal reflections are produced and sensitivity is improved. Such a crystal is schematically shown in Fig. 3.10.

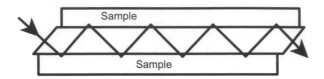

Figure 3.10 A schematic drawing of a horizontal attenuated internal reflectance (ATR). Cell showing the passage of radiation through the sample.

The factors that affect an ATR spectrum include the following:

- Wavelength of the infrared radiation
- Refractive index of the ATR crystal and sample
- Depth of penetration of the evanescent wave
- Effective path length
- Angle of incidence
- Efficiency of sample contact
- ATR crystal material

The extent in which infrared radiation extends beyond the surface of the ATR crystal and penetrates the sample depends, in part, upon the wavelength of the infrared radiation. As the wavenumber of the infrared radiation increases, the depth of penetration into the sample decreases. This results in a corresponding decrease in the band intensities of the ATR spectrum (relative to a transmission spectrum of the same sample). However, this phenomenon can be software corrected. The refractive index of the ATR crystal also influences the extent in which the evanescent wave will penetrate the sample material. The smaller the refractive index, the greater the depth of penetration. Of the two most commonly used ATR materials, ZnSe has a lower refractive index (2.4) that Ge (4), thus, ZnSe will have the greater depth penetration and will result in greater absorbance intensity.

ATR is a powerful and versatile technique. It may be used to study waste effluents, detergent systems, beverages, fermentation broths, organic solvents, household cleaners, chromatographic eluents and many other aqueous or highly absorbing solvent systems. With modifications, the ATR cell may be used to monitor components of both static and dynamically flowing systems. In the case of non-flowing viscous materials, such as gels, pastes, syrups, and greases,

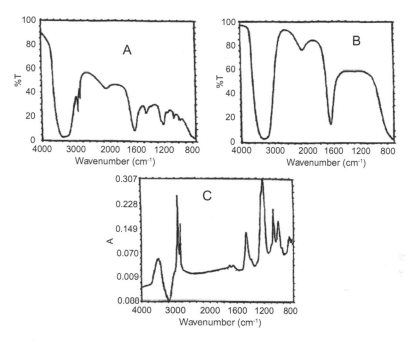

Figure 3.11 The spectrum of a commercial shampoo (A), the spectrum of water (B) and the difference spectrum (C). *Source*: Perkin–Elmer Corporation, Norwalk, CT.

the cell sits in a sampling trough containing the viscous material. As an example of the use of the ATR technique, consider the spectrum of a commercial shampoo (Fig. 3.11A); the spectrum of water (Fig. 3.11B) and the difference spectrum after water is subtracted (Fig. 3.11C).

The difference spectrum (Fig. 3.11C) clearly shows absorption bands common to the components of shampoos. These include absorptions below 1600 cm^{-1} (6.25 μ) which represent the bands of sulfonates, phosphates and alcohols. The absorptions around 2900 cm^{-1} (3.45 μ) represents the C–H stretching vibrations, also related to the components of the shampoo. The negative band near 3500 cm^{-1} (2.86 μ) is a phenomenon common to measurements of this type and has been attributed to hydrogen bonding interactions among the N–H and O–H groups of the active components and water.

3.12 Quantitative Infrared Analysis

Quantitative analysis of components in mixtures may be determined by FTIR. The procedures is software driven and generally involve either, (a) comparison of the areas under specific absorbance bands; (b) comparison of the areas under the first or second derivative of the absorbance bands; (c) comparison of the heights of absorbance bands and (d) least squares curve fitting. The latter procedure is often more accurate than the others. The "least squares curve fitting" procedure requires that a series of standard reference samples of known concentrations be prepared and their spectra obtained. The software program seeks to match one of these standard sets with the spectrum of the unknown. On this basis, the concentration of the unknown mixture is determined. All quantitative infrared procedures assume that the infrared absorbance at any frequency varies in a linear fashion with concentration (Beer's Law). The quantitative procedure works best when the analysis is conducted under identical conditions and when there is no overlapping of absorbance bands. Quantitative infrared procedures may be used to accurately analyze multicomponent systems.

3.13 Combined Thermogravimetric Analysis and FTIR (TG/FTIR)

As a background for the discussion of combined thermogravimetric analysis-FTIR, it may be appropriate to briefly describe the common thermal methods. Thermal analysis itself may be defined in a general fashion as an investigation of the effects of temperature changes on a substance. Thermoanalytical techniques are valuable for understanding the thermal properties of a variety of materials. The technique has found use in many areas such as polymer studies, explosive research, soil evaluation, environmental and mineralogical studies, metallurgical and material testing. Among the most common thermal techniques are thermogravimetry (TG), differential thermal analysis (DTA), derivative thermogravimetric analysis (DTG) and differential scanning calorimetry (DSC).

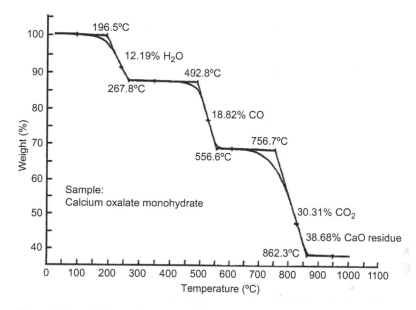

Figure 3.12 A TG curve showing the thermal decomposition of calcium oxalate monohydrate.

TG measures the weight loss of the sample as it is heated under controlled conditions of temperature and pressure (Fig. 3.12). These studies may be conducted under normal, increased or reduced pressures and in different gas atmospheres. Valuable information relating to the thermal stability and composition of the sample material may be obtained from these studies.

DTA compares the temperature changes in the sample verses an inert material as both materials are heated (or cooled) at a uniform rate. Temperature changes in a sample may be ether endothermic or exothermic, depending upon the nature of the physical or chemical changes involved. Typical among these changes include melting, boiling, changes in crystalline structure, oxidation, reduction, decomposition, phase transitions and dehydration. Endothermic effects usually accompany phase transitions, dehydrations, reductions and certain decomposition reactions. Among the common exothermic processes are oxidation and crystallization.

Derivative thermogravimetric analysis (DTG) provides the first derivative of the mass curve (dm/dt). In other words, the DTG curve

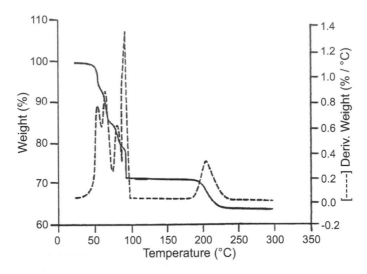

Figure 3.13 A comparison of the high-resolution TG curve (solid line) and DTG curve (dashed line) of cupric sulfate pentahydrate ($CuSO_4 5H_2O$). *Source*: The *TA Hotline*, Volume 2, 1991, TA Instrument, Inc., New Castle, DE.

is the TG curve expressed in a different visual form. Information obtained from the DTG curve may be used to extract kinetic information relating to the thermal decomposition of the sample material. It is also useful in resolving subtle thermal phenomena that may not be obvious from a study of the TG curve (compare the curves in Fig. 3.13).

Differential scanning calorimetry (DSC) measures heat flow in or out of the sample (as opposed to temperature differences between the sample and an inert reference, as in DTA). In DSC, the sample is heated or cooled at a linear rate while the temperature of the sample and reference is maintained isothermally with respect to each other by the application of electrical energy. The amount of heat energy necessary to maintain the sample isothermally is recorded on the DSC curve. The resulting curve (Fig. 3.14) is similar in appearance to the DTA curve, but it measures the heat flow rate of the sample with respect to temperature (dH/dt). The area under both the DTA and DSC curve is proportional to the change in enthalpy (ΔH).

Despite the valuable information obtained from the briefly described thermal methods, none of them can provide information

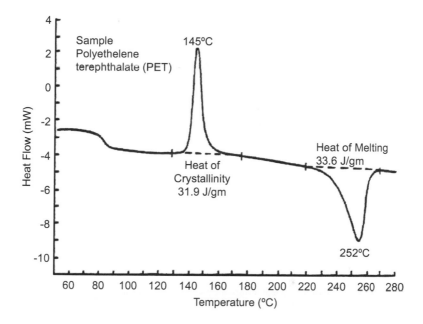

Figure 3.14 A Typical DSC curve.

on the structural identity of decomposition products. Yet this information is vital to the complete characterization of material behavior under thermal conditions. By coupling thermal systems with Fourier Transform infrared spectrometers (or mass spectrometers) this limitation is removed. The combined system provides more information than the systems operating separately. With the aid of sophisticated FTIR and thermal analysis software, commercial FTIR/TG systems can provide an abundance of information relating to the thermal decompositions of materials. Some of these techniques are discussed below.

Total (or Full) FTIR Evolved Gas Thermogram: This technique provides FTIR spectral information on all evolved gases as a function of time. Consider, for example, the total FTIR gas thermogram for the thermal decomposition of vinyl acetate (13%) vinyl chloride co-polymer (Fig. 3.15a), the specific thermogram of acetic acid (3.15b) and carbon dioxide (3.15c). Acetic acid and carbon dioxide are products of the thermal decomposition.

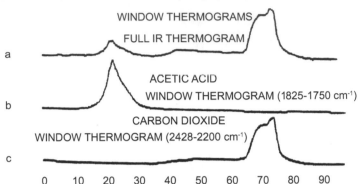

TG/FT-IR ANALYSIS VINYL ACETATE (15%) VINYL CHLORIDE COPOLYMER

Figure 3.15 Examples of IR thermograms of vinyl chloride (13%)–poly vinyl acetate (PVC/PVA) co-polymer showing the total gas thermogram (a), the specific frequency thermograms of acetic acid (b) and carbon dioxide (c). *Source*: Cassel, B. and G. McClure, Am. Lab., January, 1989.

Specific Gas Profile: In this profile a specific frequency region is scanned over time. This profile is useful for studying the evolution of specific gases. For instance, the evolution of HBr over a specific time frame may be observed by monitoring the H–Br stretching frequency.

Specific Frequency Thermogram (or Functional Group Profile): Here, the absorbance within a small frequency region is monitored as a function of time or temperature. This evolved gas profile is useful in monitoring the evolution of gases with common functional groups. Consider as an example, Fig. 3.15, which shows the evolution of acetic acid and carbon dioxide, products resulting from the thermal decomposition of vinyl chloride/polyvinyl acetate (PVC/PVA) co-polymer.

3.14 The TG/FTIR Interface

While the FTIR/TG interface may vary among manufacturer, they all embody the same general principle. Most of the commercial interfaces consist of an inert, temperature controlled, heated transfer line. This line allows the gaseous components to enter a heated gas cell (light pipe) located on the optical bench of the FTIR. As the decomposition gases enter the heated light pipe, they encounter infrared radiation and their spectra are determined. A schematic illustration of the light pipe interface and a photograph of a combined TG/FTIR are shown in Figs. 3.16 and 3.17, respectively.

3.15 GC/MS, GC/FTIR and GC/MS/FTIR

The coupling of a gas chromatograph with a mass spectrometer and/or Fourier Transform Infrared spectrophotometer is significant in that it allows the components of complex system to be separated and identified. The sensitivity of these systems is such that components, with concentrations in the nanogram range, can be separated, their spectra obtained and their structures determined. The complementary structural information provided by simultaneous GC/MS/FTIR systems enhances component identification and is the principal advantage of GC/MS/FTIR over GC/MS or GC/FTIR. As early as 1968, the analytical advantages of a combined GC/MS/FTIR system were discussed, but computer technology and software programming had not been sufficiently developed to make such systems practical. In the intervening years many developments took place in this area, but it was not until 1980 that a combined GC/MS/FTIR was developed. Presently, GC/MS/FTIR systems may be purchased for less than $100,000, placing these sophisticated systems within financial reach of some small colleges and universities. These systems are capable of rapidly analyzing complex mixtures from one injection into the GC.

Some manufacturers have developed combined GC/MS/FTIR systems configured in a parallel fashion. In this configuration, the GC effluent is divided, with a portion directed to the MS and the remaining portion directed to the FTIR. The serial configuration has the effluent passing through an infrared detector light pipe (Fig. 3.16) and from there to the mass spectrometer. One manufacturer has the transfer line to and from the IR light pipe and the transfer line to the mass spectrometer all located in the oven of the GC. According to the manufacturer, this arrangement allows the system to be configured in either a parallel or a serial manner.

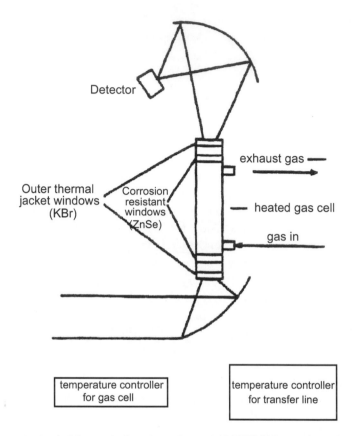

Figure 3.16 A Schematic drawing of a typical FTIR/TG interface. *Source*: Biorad, Digilab Division, Cambridge, MA.

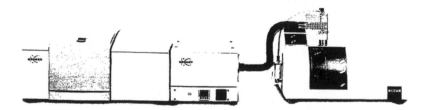

Figure 3.17 A combined FTIR/TG system. *Source*: Bruker Corporation.

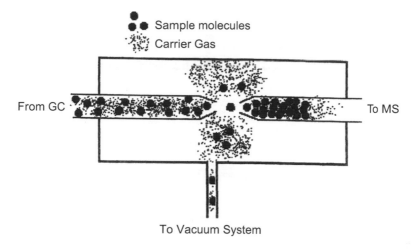

Figure 3.18 The jet spray.

3.16 The GC/MS and GC/FTIR Interfaces

The coupling of a GC and MS must be designed to allow the components in the GC effluent to enter the chamber of the mass spectrometer without appreciably changing its vacuum. This is especially true for packed columns that are characterized by large amounts of carrier gases. This requirement can be accomplished with a jet spray, which is designed to partially remove the carrier gas (usually helium) before the GC effluent enters the chamber of the mass spectrometer. The jet spray illustrated in Fig. 3.18 is surrounded by a vacuum which removes a significant portion of the carrier gas as the GC effluent passes through a small orifice into the chamber of the mass spectrometer. After passing through the orifice, most of the carrier gas diffuses away (since it has a higher diffusion rate that the sample components). Studies have shown that the jet spray device can remove about 90 percent of the carrier gas.

In the combined capillary GC/MS/FTIR systems, the capillary column of the GC is usually placed directly into the FTIR light pipe. The spectra of the gases in the light pipe are obtained and from there the GC effluent travels to the mass spectrometer through a smaller

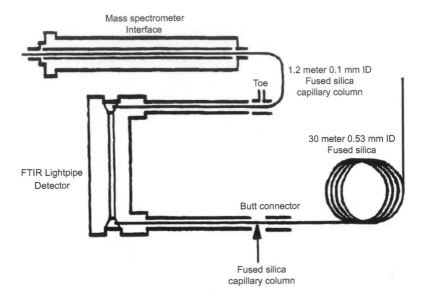

Figure 3.19 Schematic diagram of a GC/MS/FTIR interface. *Source*: Leibrand, R. J., Am. Lab., December, 1988.

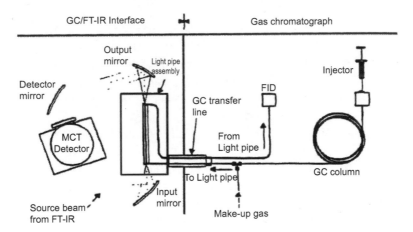

Figure 3.20 A schematic of GC/FTIR Interface. *Source*: Nicolet Instrument Corporation, Madison, WI.

capillary tube. A typical GC/MS/FTIR interface is shown in Fig. 3.19, and in Fig. 3.20 are shown the schematics of GC/FTIR interfaces.

Chapter 4

Problems in Infrared Spectroscopy

(1) Convert 2526 cm^{-1} to micrometers (μm).

(2) Briefly explain the theory of infrared spectroscopy.

(3) Calculate the approximate stretching frequency of the C–F bond which has a force constant of 5.6×10^5 dynes/cm.

(4) How many fundamental vibrations are predicted for the non-linear water molecule?

(5) Name four reasons why all predicted fundamental vibrations of a bond may not appear in the infrared spectrum.

(6) Discuss the advantages of FTIR over dispersive infrared spectroscopy.

(7) Explain why the carbonyl stretching frequency is always strong.

(8) A certain compound gives carbonyl absorption at approximately 1775 cm^{-1}, which of the structures shown below might it be? Explain.

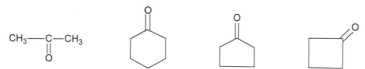

$$CH_3\!-\!\underset{\underset{O}{\|}}{C}\!-\!CH_3$$

(9) When a carbonyl group is conjugated with a C=C double bond, the observed carbonyl frequency is lowered by about

Infrared Spectroscopy
James M. Thompson
Copyright © 2018 Pan Stanford Publishing Pte. Ltd.
ISBN 978-981-4774-78-9 (Hardcover), 978-1-351-20603-7 (eBook)
www.panstanford.com

20–30 cm^{-1} as shown below. In terms of resonance interaction, suggest a plausible explanation for this frequency shift.

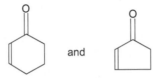

1715 cm-1   1685 cm-1

(10) Using the information in problem 9 and based on a carbonyl absorption of 1715 cm^{-1} (5.83 μm) for cyclohexanone and 1745 cm^{-1} (5.73 μm) for cyclopentanone, approximate the carbonyl absorption for the structures shown below.

and

(11) The infrared spectrum of a certain compound shows two carbonyl absorptions. With this limited information, suggest a probable structural class for the compound.

(12) A certain nitrile shows a medium C≡N stretching vibration near 4.45 μm. Convert this value to wavenumbers in cm^{-1} and frequency in sec^{-1} and Hertz.

(13) The C-H olefinic stretching vibration of 1-heptene occurs at 3096 cm^{-1} (3.23 μm). What is the energy in ergs that corresponds to a photon of infrared radiation having this wavelength?

(14) From the C≡N stretching vibration in problem 12, calculate the approximate force constant for this bond.

(15) The O–H stretching absorption of the compound below does not change to a great extent with changes in concentration in carbon tetrachloride solution. Explain.

(16) Explain how the infrared spectra of the following structures might differ in the carbonyl region.

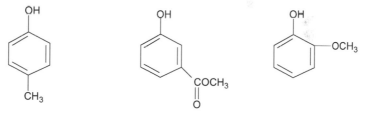

(17) Of the three compounds shown below, which one would probably show the sharpest O–H stretching vibration? Explain.

(18) Discuss how the infrared spectra of non-branched and branched hydrocarbons will differ.

(19) What are the structure implications of a compound that shows an absorption above 3000 cm^{-1} (below 3.33 µm)?

(20) How could one use infrared spectroscopy to distinguish between the pairs of compounds that follow? Indicate the characteristic absorptions that you would expect for each structure pair.

(a) 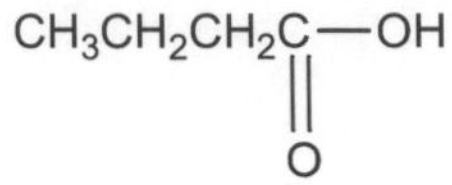 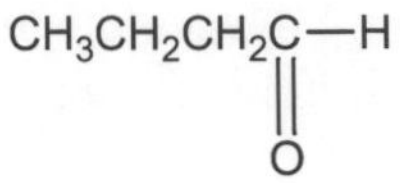 $CH_3CH_2CH_2\overset{\overset{\displaystyle O}{\|}}{C}-OH$ $\qquad$ $CH_3CH_2CH_2\overset{\overset{\displaystyle O}{\|}}{C}-H$

(b) 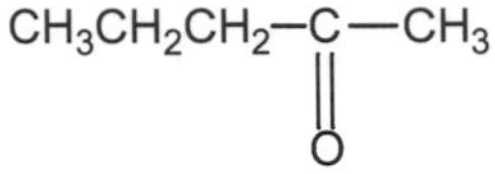 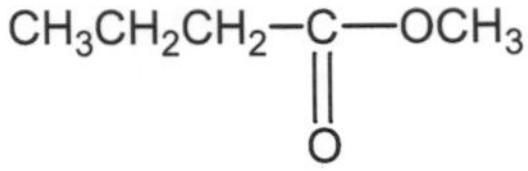 $CH_3CH_2CH_2-\overset{\overset{\displaystyle O}{\|}}{C}-CH_3$ $\qquad$ $CH_3CH_2CH_2-\overset{\overset{\displaystyle O}{\|}}{C}-OCH_3$

(c)

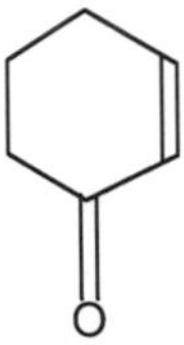

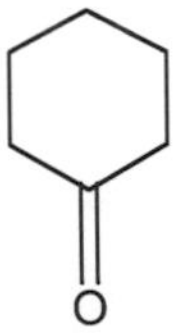

(d) $CH_3CH_2C\equiv CH$ $\qquad$ $CH_3CH_2C\equiv CCH_3$

(e) $CH_3C\equiv N$ $\qquad$ $CH_3C\equiv CH$

(f)

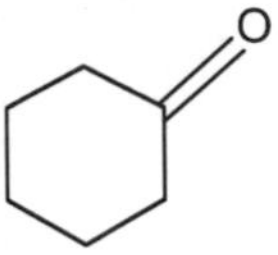

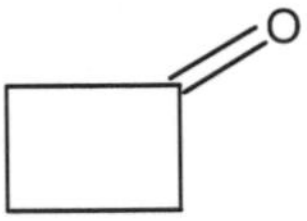

(g) $CH_3CH_2CH_2CH_2CH_3$ $\qquad$ $CH_3\overset{\overset{\displaystyle CH_3}{|}}{\underset{\underset{\displaystyle CH_3}{|}}{C}H}CHCH_2CH_3$

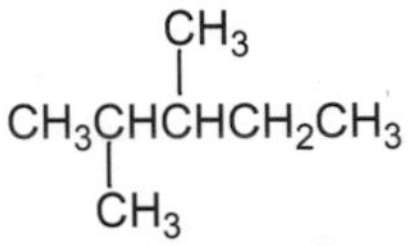

(h) $CH_3CH_2CH_2CH_2CH_2\overset{\underset{\displaystyle H}{|}}{C}=CH_2$ $\qquad$ $CH_3CH_2CH_2CH_2CH_2C\equiv CH$

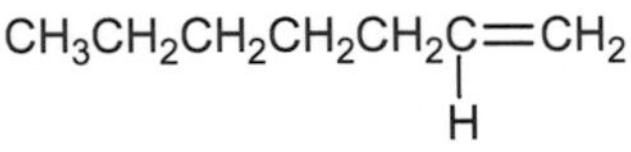

(i) 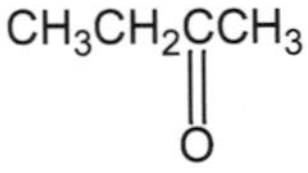 $CH_3CH_2\overset{\overset{\displaystyle O}{\|}}{C}CH_3$

$CH_3CH_2CH_2OCH_3$

(j) $CH_3CH_2CH_2OH$ $\qquad$ $-CH_2COOH$

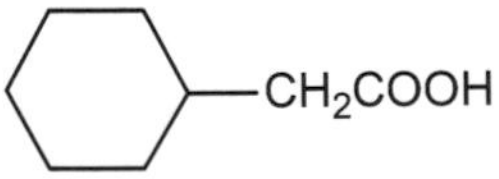

(k) 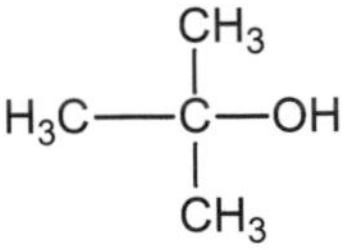 $H_3C-\overset{\overset{\displaystyle CH_3}{|}}{\underset{\underset{\displaystyle CH_3}{|}}{C}}-OH$ $\qquad$ $CH_3CH_2CH_2OH$

(21) From a consideration of the three spectra below, identify the alcohol, ester and acid.

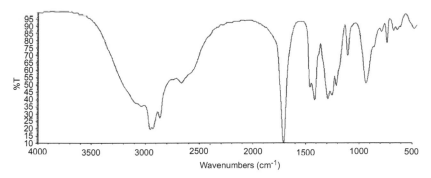

Figure 4.1 Solution in CCL₄.

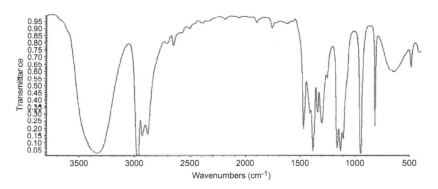

Figure 4.2 Solution in CCL₄.

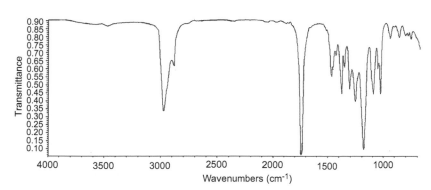

Figure 4.3 Neat.

(22) Using the chemical formula (listed below each spectrum) determine the structures of the compounds represented by the spectra in Figs. 4.4–4.32.

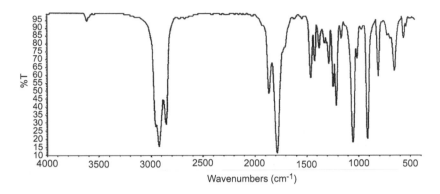

Figure 4.4 $(C_4H_4O_3)$.

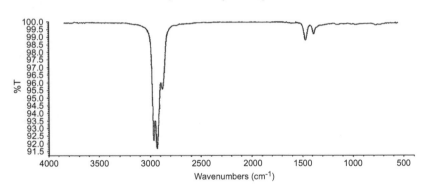

Figure 4.5 (C_9H_{20}).

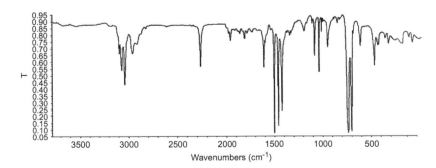

Figure 4.6 (C_8H_7N).

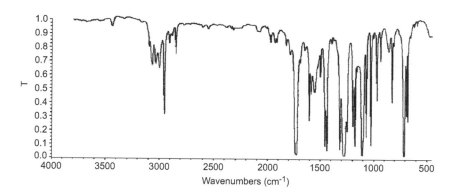

Figure 4.7 $(C_8H_8O_2)$.

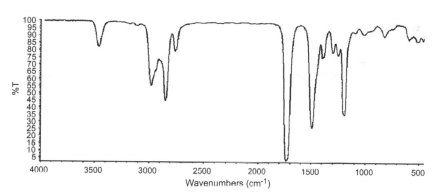

Figure 4.8 (C_3H_7NO).

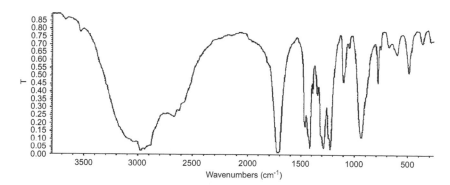

Figure 4.9 $(C_4H_8O_2)$.

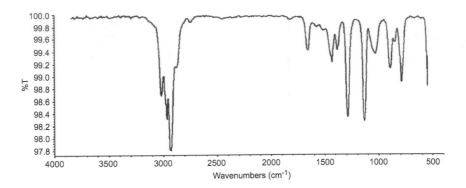

Figure 4.10 (C_4H_7Br).

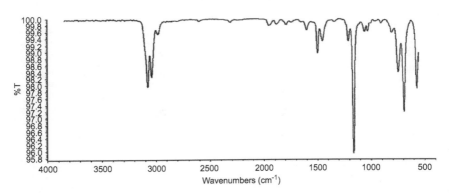

Figure 4.11 (C_7H_7I).

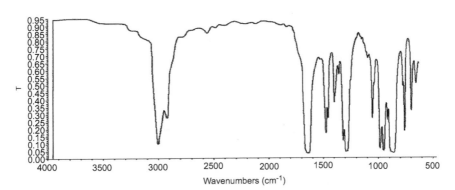

Figure 4.12 $(C_3H_7O_3N)$.

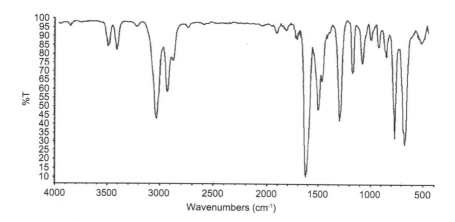

Figure 4.13 (C_7H_9N).

The compound represented by the IR spectrum in Fig. 4.14 has the formula C_7H_{14}. It decolorizes bromine in carbon tetrachloride and when reacted with ozone, formaldehyde is formed as one of the products. What is the probable structure of the compound?

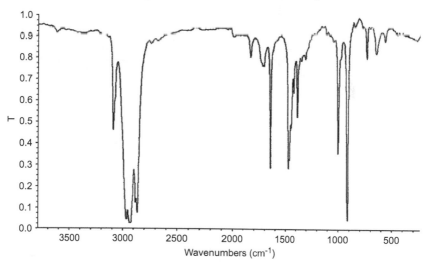

Figure 4.14 (C_7H_{14}).

The compound represented by the spectrum shown in Fig. 4.15 is used in the preparation of a well-known polymeric material.

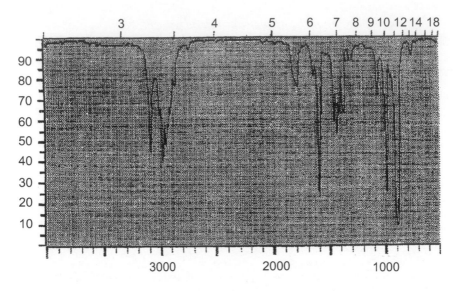

Figure 4.15 (C_5H_8).

The IR spectrum of the compound shown in Fig. 4.16 has the formula (C_4H_8O). It reacts with 2,4-DNP to give a precipitate and also gives a precipitate when reacted with ammoniacal silver nitrate. What is the probable structure of this compound?

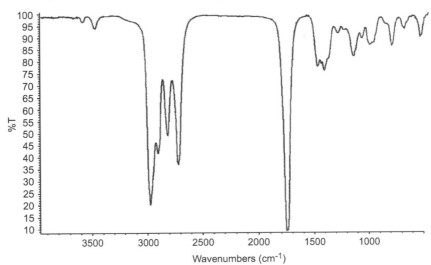

Figure 4.16 (C_4H_8O).

When the compound represented by Fig. 4.17 was heated with P_4H_{10}, a compound with the formula C_2H_3N was formed. What is the probable structure of this compound?

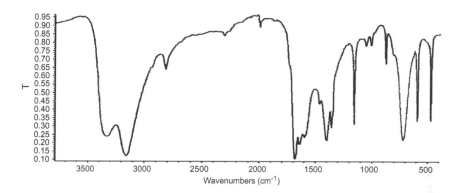

Figure 4.17 (C_2H_5ON).

When the compound represented by Fig. 4.18 was heated, it produced a cyclic imide. What is the probable structure of the compound?

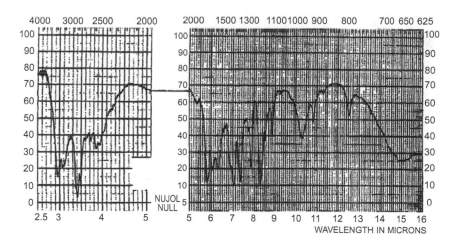

Figure 4.18 ($C_4H_7O_3N$).

When the compound represented by Fig. 4.19 is hydrolyzed, acetic acid is one of the products. What is the probable structure of the compound?

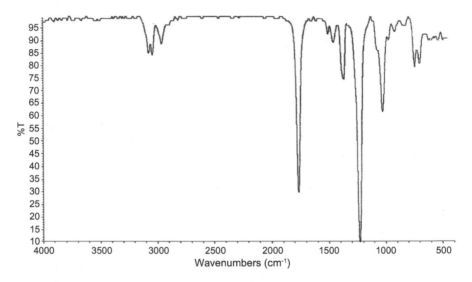

Figure 4.19 $(C_9H_{10}O_2)$.

In determining the structure of the compound represented by the spectrum in Fig. 4.20, study the carbonyl region.

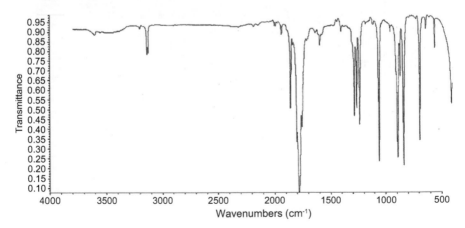

Figure 4.20 $(C_4H_2O_3)$.

Of the two spectra represented by Figs. 4.21a and 4.21b, one represents a primary alcohol and the other a secondary alcohol. Label each spectrum and give a rationale for your selection.

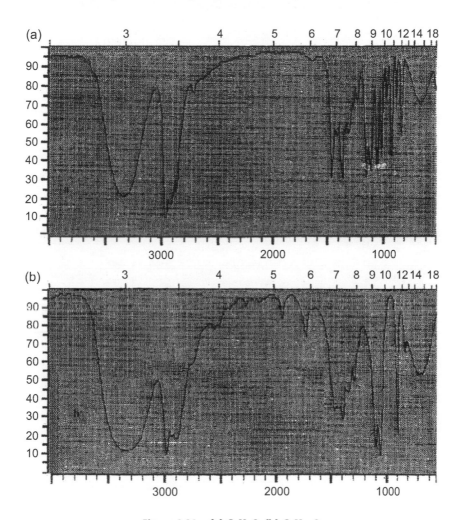

Figure 4.21 (a) C_2H_6O, (b) $C_6H_{14}O$.

The compounds represented by Figs. 4.22a and 4.22b are substituted pentanes with the formula C_7H_{16}. Determine the structures of each of the compounds and give a rationale for your answer.

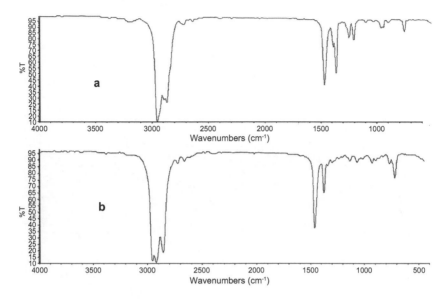

Figure 4.22

Identify the glycol in Fig. 4.23.

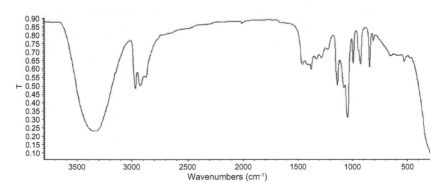

Figure 4.23 $(C_3H_8O_2)$.

The two compounds represented by the spectra shown in Figs. 4.24a and 4.24b both have the formula $C_4H_{10}O$. Only compound 4.24a gives a positive iodoform test. Match each spectrum with its probable structure.

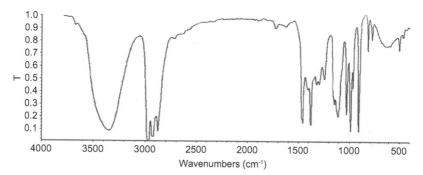

Figure 4.24 a

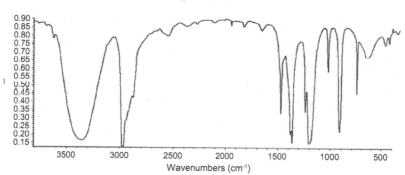

Figure 4.24 b

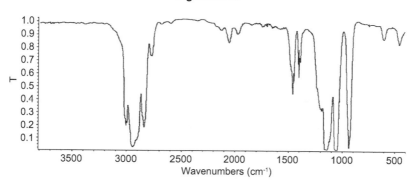

Figure 4.24 c ($C_3H_8O_2$).

The compound represented by Fig. 4.25 is a saturated hydrocarbon. Does the compound have more than four carbon atoms? Explain.

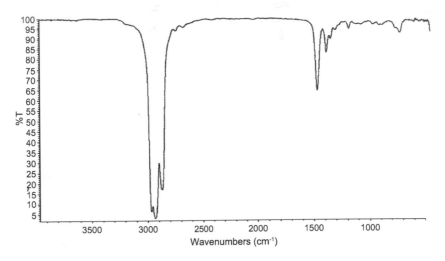

Figure 4.25

In Fig. 4.26, the structure is an aromatic ether.

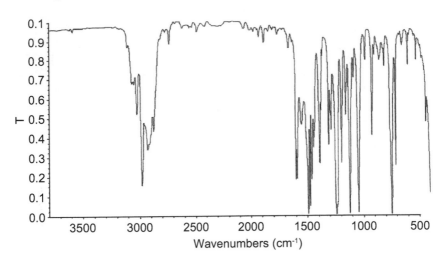

Figure 4.26 $(C_9H_{12}O)$.

The compounds represented by the IR spectra in Figs. 4.27a and 4.27b are aromatic anhydrides. Which spectrum represents the cyclic anhydride? Explain.

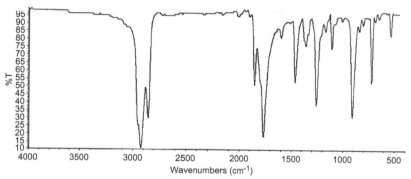

Figure 4.27 a

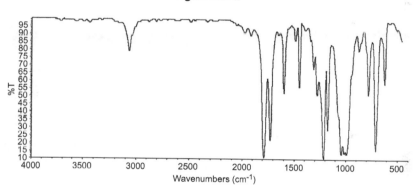

Figure 4.27 b

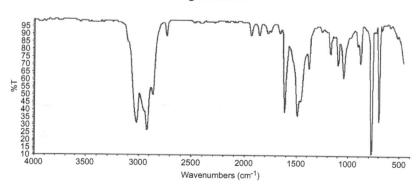

Figure 4.28 (C_8H_{10}).

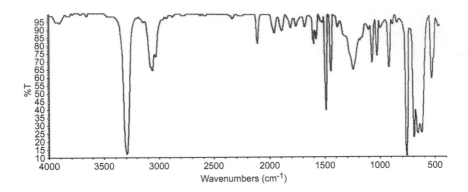

Figure 4.29 (C_8H_6).

The compound represented by Fig. 4.30 adds two molar equivalent of hydrogen. What is the probable structure of the compound?

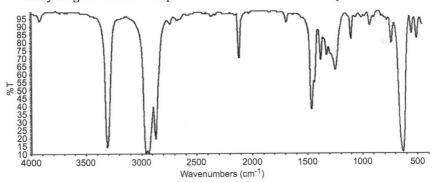

Figure 4.30 (C_6H_{10}).

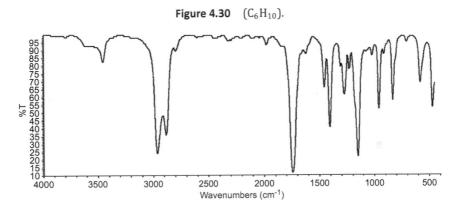

Figure 4.31 (C_5H_8O).

The compound represented by Fig. 4.32 has a foul odor. What is the probable structure of the compound?

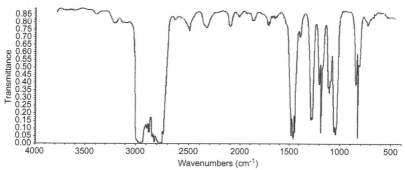

Figure 4.32 (C_3H_9N), gas phase.

(23) Match the names below with the spectra that follow.

Undecane	Furan
Anisole	Methyl acetate
Benzene	Thiophenol
Catechol	Acrolein
Cycloheptanone	N-Methylacetamide
Deoxybenzoin	4-Methyl-4-heptanol
Benzyl phenyl acetate	p-Dibromobenzene
Allyl acetate	Methyl benzoate
Adipoyl chloride	Dimethylsulfone
Aniline	Dimethoxymethylphenyl silane
Benzoin	

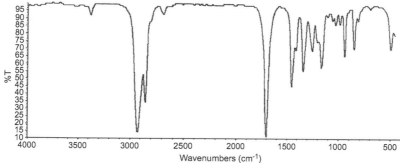

Figure 4.33

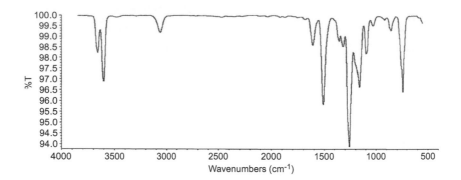

Figure 4.34

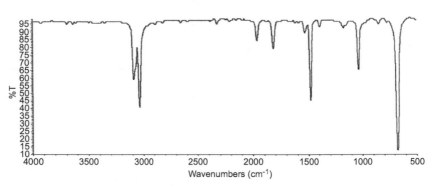

Figure 4.35

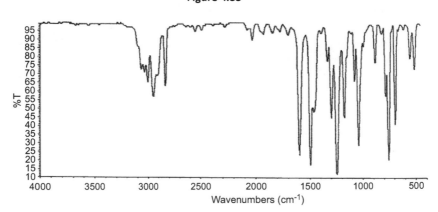

Figure 4.36

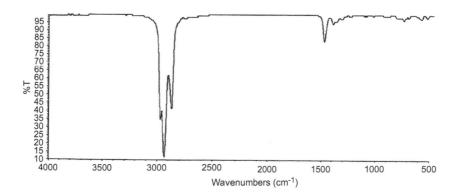

Figure 4.37

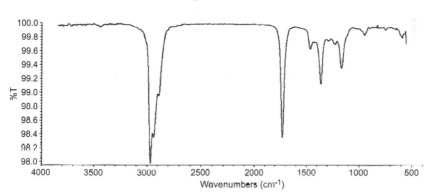

Figure 4.38

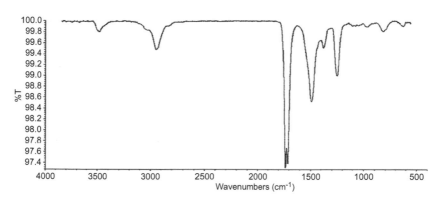

Figure 4.39

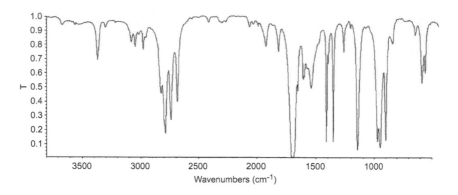

Figure 4.40

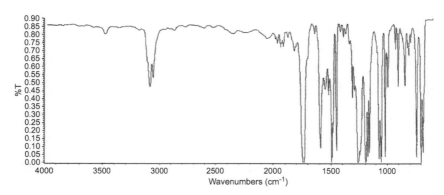

Figure 4.41

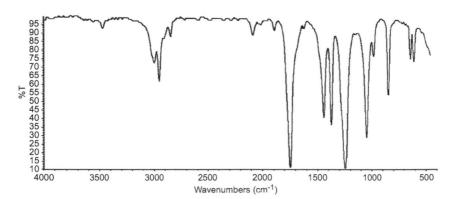

Figure 4.42

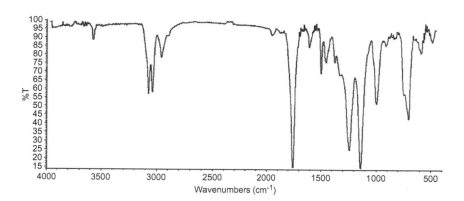

Figure 4.43

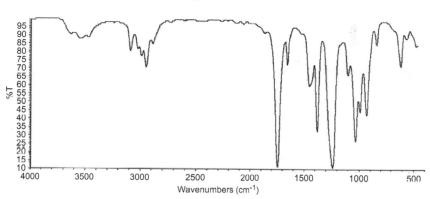

Figure 4.44

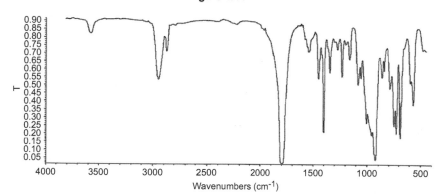

Figure 4.45

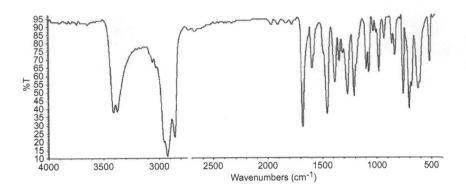

Figure 4.46

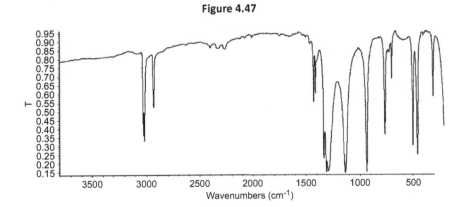

Figure 4.47

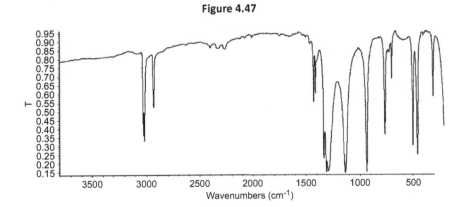

Figure 4.48

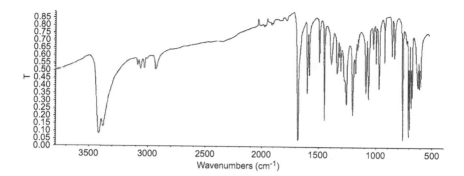

Figure 4.49

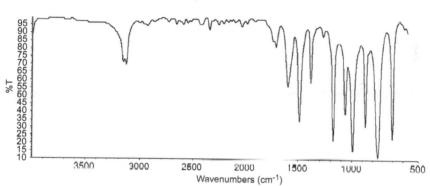

Figure 4.50

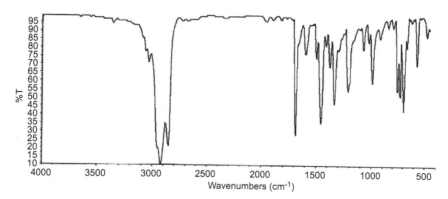

Figure 4.51

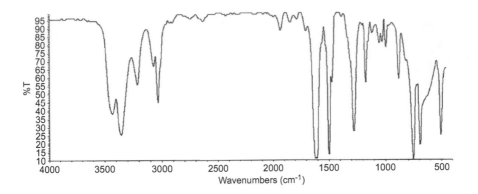

Figure 4.52

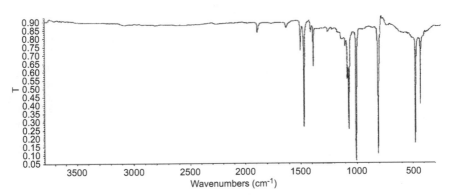

Figure 4.53

(24) In the examples that follow, assign the three structures to the three spectra below and justify your assignments based on bond frequencies.

a   b   c

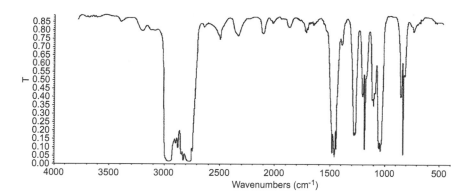

Figure 4.54

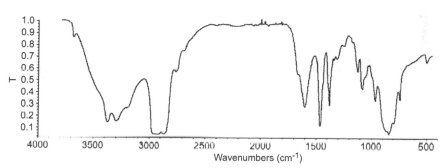

Figure 4.55

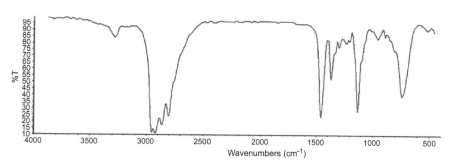

Figure 4.56

Assign the infrared spectra below to the structures and give an explanation for your assignments.

$CH_3CH_2CH_2CH_2-\overset{\displaystyle O}{\underset{\displaystyle \|}{C}}-CH_3$

a

$CH_3(CH_2)_8COOH$

c

$CH_3CH_2CH_2CH_2OCH_2CH_2CH_2CH_3$

b

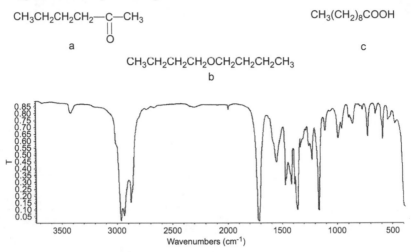

Figure 4.57

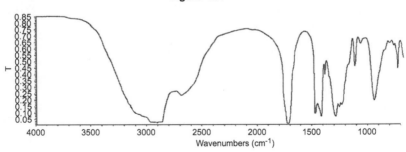

Figure 4.58

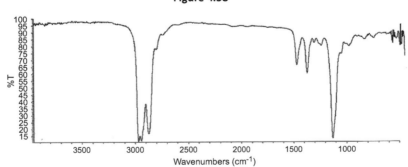

Figure 4.59

Assign the infrared spectra below to the structures and give an explanation for your assignments.

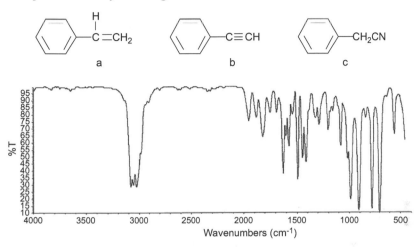

a   b   c

Figure 4.60

Figure 4.61

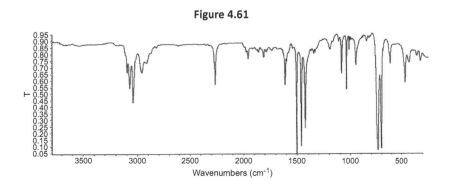

Figure 4.62

Assign the infrared spectra below to the structures and give an explanation for your assignments.

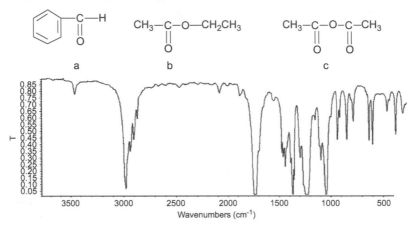

a   b   c

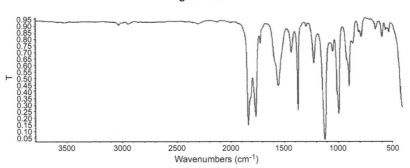

Figure 4.63

Figure 4.64

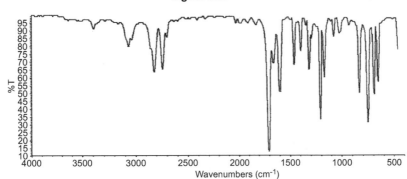

Figure 4.65

Assign the infrared spectra below to the structures and give an explanation for your assignments.

CH$_3$CH$_2$CH$_2$CH$_2$CH$_2$COOH   H$_2$C=CCH$_2$OCH$_2$C—CH$_2$ CH$_3$CCH$_2$CH$_3$

a   b   c

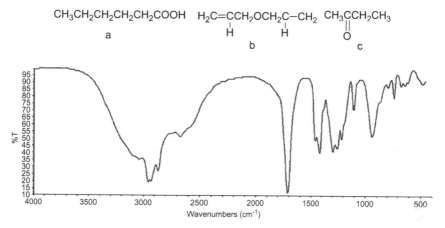

Figure 4.66

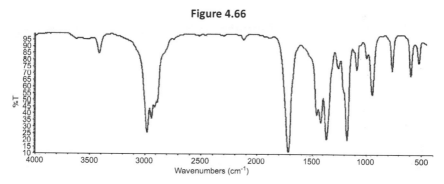

Figure 4.67

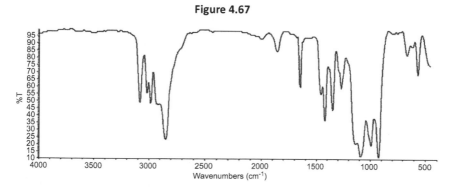

Figure 4.68

Assign the infrared spectra below to the structures and give an explanation for your assignments.

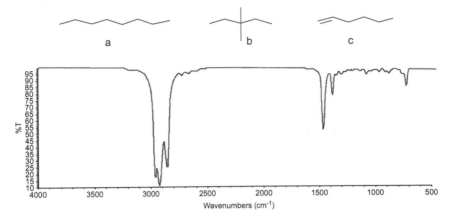

a   b   c

Figure 4.69

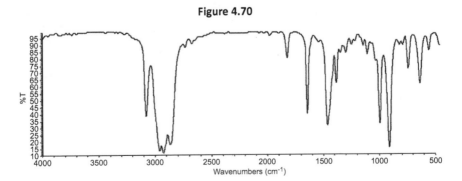

Figure 4.70

Figure 4.71

Assign the infrared spectra below to the structures and give an explanation for your assignments.

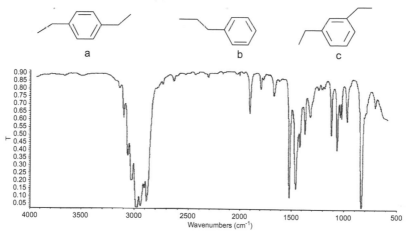

a   b   c

Figure 4.72

Figure 4.73

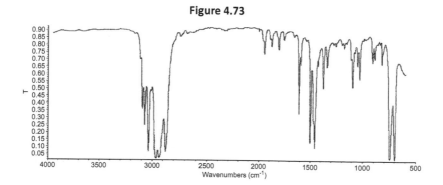

Figure 4.74

Assign the infrared spectra below to the structures and give an explanation for your assignments.

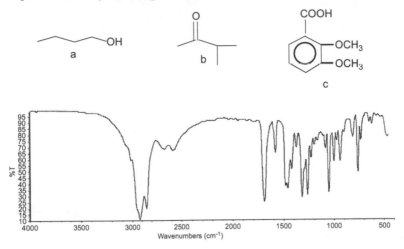

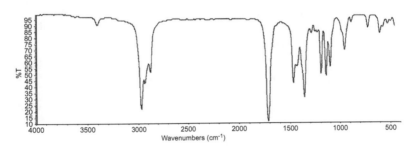

Figure 4.75

Figure 4.76

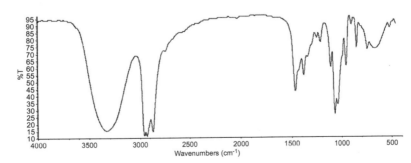

Figure 4.77

Assign the infrared spectra below to the structures, and give an explanation for your assignments.

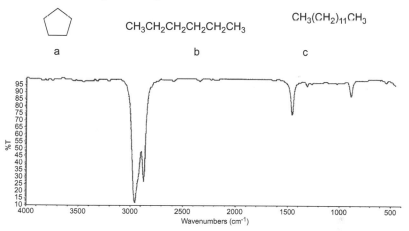

a

$CH_3CH_2CH_2CH_2CH_2CH_3$

b

$CH_3(CH_2)_{11}CH_3$

c

Figure 4.78

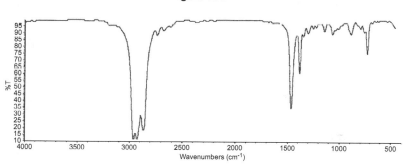

Figure 4.79

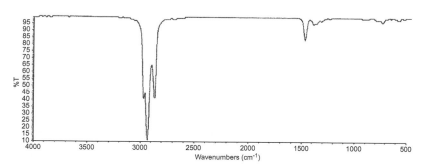

Figure 4.80

Assign the infrared spectra below to the structures and give an explanation for your assignments.

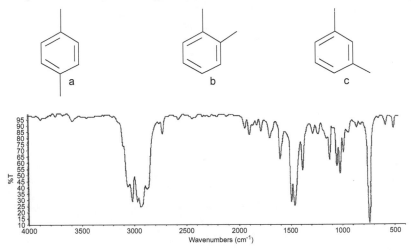

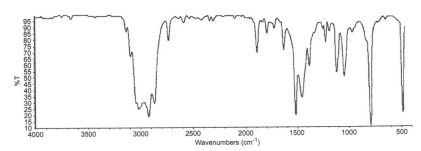

Figure 4.81

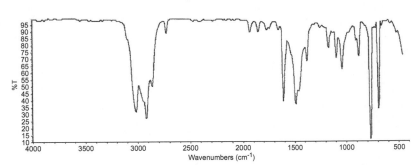

Figure 4.82

Figure 4.83

Assign the infrared spectra below to the structures and give an explanation for your assignments.

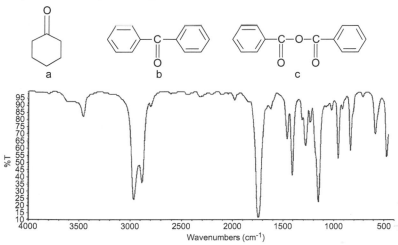

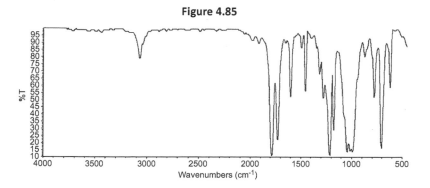

Figure 4.84

Figure 4.85

Figure 4.86

Assign the infrared spectra below to the structures and give an explanation for your assignments.

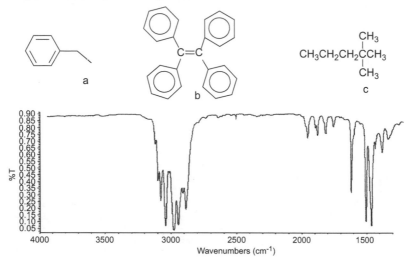

Figure 4.87

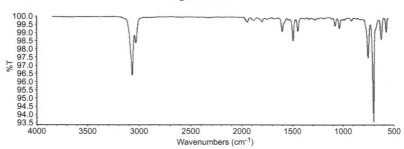

Figure 4.88

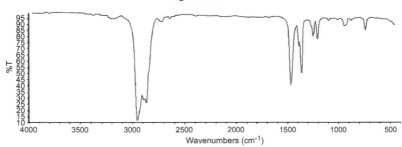

Figure 4.89

Assign the infrared spectra below to the structures, and give an explanation for your assignments.

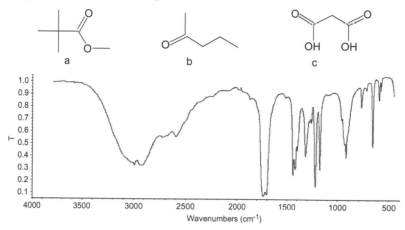

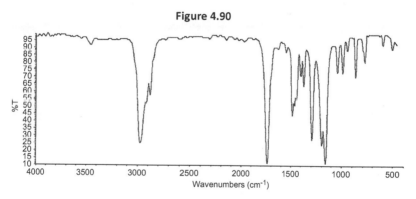

Figure 4.90

Figure 4.91

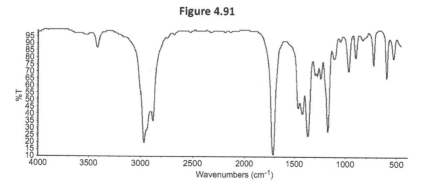

Figure 4.92

Assign the infrared spectra below to the structures, and give an explanation for your assignments.

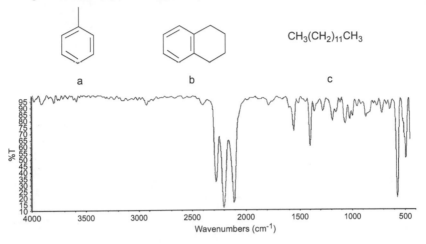

a   b $CH_3(CH_2)_{11}CH_3$

c

Figure 4.93

Figure 4.94

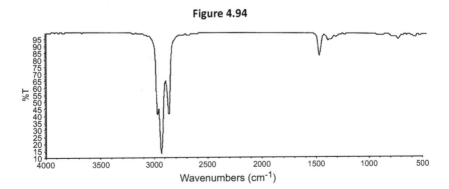

Figure 4.95

Assign the infrared spectra below to the structures, and give an explanation for your assignments.

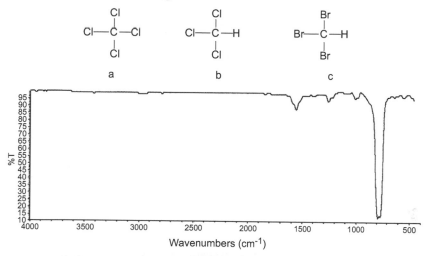

a   b   c

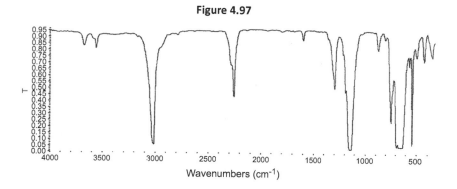

Figure 4.96

Figure 4.97

Figure 4.98

Assign the infrared spectra below to the structures, and give an explanation for your assignments.

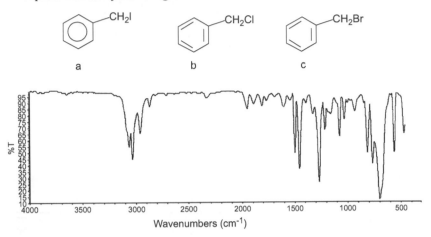

Figure 4.99

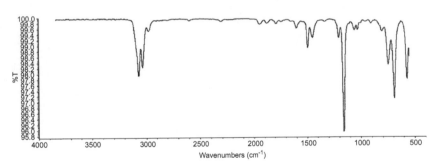

Figure 4.100

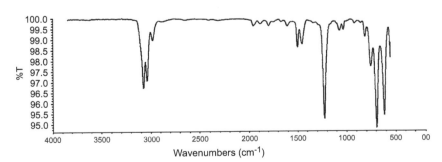

Figure 4.101

Assign the infrared spectra below to the structures, and give an explanation for your assignments.

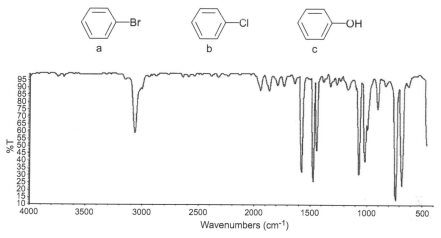

a   b   c

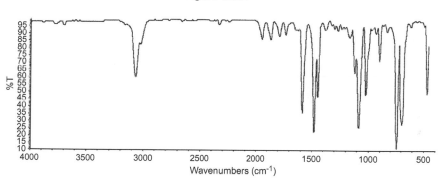

Figure 4.102

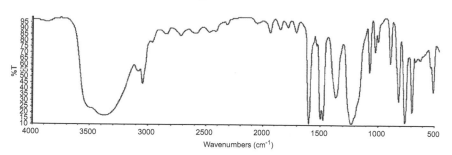

Figure 4.103

Figure 4.104

Answers to Selected Questions

Problem 21

Figure 4.1	Hexanoic acid
Figure 4.2	2-Propanol
Figure 4.3	Ethyl butanoate

Problem 22

Figure 4.4	Succinic anhydride
Figure 4.5	3-Methyl octane
Figure 4.6	Benzene acetonitrile
Figure 4.7	Methylbenzoate
Figure 4.8	N-Ethylformamide
Figure 4.9	Butanoic acid
Figure 4.10	2-Bromo-2-butene
Figure 4.11	Benzyl iodide
Figure 4.12	Propyl nitrate
Figure 4.13	3-Methylbenzylamine
Figure 4.14	1-Heptene
Figure 4.15	Isoprene
Figure 4.16	Butanal
Figure 4.17	Acetamide
Figure 4.18	Succinamic acid
Figure 4.19	Benzyl acetate
Figure 4.20	Maleic anhydride
Figure 4.21a	Ethyl alcohol
Figure 4.21b	4-Methyl-1-pentanol
Figure 4.22a	2,2-Dimethylpentane
Figure 4.22b	Heptane
Figure 4.23	1,2-Propanediol
Figure 4.24a	2-Butanol
Figure 4.24b	t-Butyl alcohol
Figure 4.24c	Dimethoxymethane
Figure 4.25	Octane
Figure 4.26	o-Methylphenetole
Figure 4.27a	Phthalic anhydride

Figure 4.27b   Benzoic anhydride
Figure 4.28   m-Xylene
Figure 4.29   Phenylacetylene
Figure 4.30   1-Hexyne
Figure 4.31   Cyclopentanone
Figure 4.32   Trimethylamine
Figure 4.33   Cycloheptanone
Figure 4.34   Catechol
Figure 4.35   Benzene
Figure 4.36   Anisole
Figure 4.37   Undecane
Figure 4.38   4-Methyl-2-heptanone
Figure 4.39   N-Methylacetamide
Figure 4 40   Acrolein
Figure 4.41   Phenyl benzoate
Figure 4.42   Methyl acetate
Figure 4.43   Benzyl phenyl acetate
Figure 4.44   Allyl acetate
Figure 4.45   Adipoyl Chloride
Figure 4.46   Benzoin
Figure 4.47   Methyl benzoate
Figure 4.48   Dimethyl sulfone
Figure 4.49   Dimethoxymethylphenyl silane
Figure 4.50   Furan
Figure 4.51   Deoxybenzoin
Figure 4.52   Aniline
Figure 4.53   p-Dibromobenzene
Figure 4.54   Trimethylamine
Figure 4.55   Butylamine
Figure 4.56   Dibutylamine
Figure 4.57   2-Hexanone
Figure 4.58   Decanoic acid
Figure 4.59   Butyl ether
Figure 4.60   Styrene (98%)
Figure 4.61   Phenyl acetylene
Figure 4.62   Benzylnitrile
Figure 4.63   Ethylacetate

Figure 4.100   Benzyl iodide
Figure 4.101   Benzyl bromide
Figure 4.102   Bromobenzene
Figure 4.103   Chlorobenzene
Figure 4.104   Phenol

Suggested Readings and References

Books

1. Silverstein R. M., G. C. Bassler and T. C. Morrill, *Spectrometric Identification of Organic Compounds*, 5th ed., New York, John Wiley (1991).
2. Schrader B., ed., *Infrared and Raman Spectroscopy: Methods and Applications*, Weinheim, New York, VCH (1994).
3. Roeges N. P. G., *A Guide for the Complete Interpretation of Infrared Spectra of Organic Structures*, New York, John Wiley & Sons (1994).
4. Coleman P. B., ed., *Practical Sampling Techniques for Infrared Analysis*, Boca Raton, Fl, CRC Press (1993).
5. Twardowski J., *Raman and IR Spectroscopy in Biology*, New York, Ellis Horwood (1992).
6. Johnston S., *Fourier Transform Infrared: A Constantly Evolving Technology*, New York, Ellis Horwood (1991).
7. Colthup N. B., *Introduction to Infrared and Raman Spectroscopy*, 3rd ed., Boston, Academic Press (1990).
8. White R., *Chromatography/Fourier Transform Infrared Spectroscopy and Its Applications*, New York, M. Dekker (1990).
9. Herres W., *HRGC-FTIR-Capillary Gas Chromatography-Fourier Transform Infrared Spectroscopy, Theory and Applications*, Heidelberg, New York, A. Huthig Verlag (1987).
10. *Advances in Applied Fourier Transform Infrared Spectroscopy*, New York, John Wiley & Sons (1988).
11. *Infrared Microspectroscopy, Theory and Applications*, New York, H. Dekker (1988).
12. *The Design, Sample Handling, and Applications of Infrared Microscopes*, A Symposium, Philadelphia, Pa., ASTM (1987).

13. Symposium on Fourier Transform Infrared Characterization of Polymers (1984, Philadelphia, Pa.), *Fourier Transform Infrared Characterization of Polymers*, New York, Plenum Press (1987).

14. *Laboratory Methods in Vibrational Spectroscopy*, 3rd ed. Chichester; New York, John Wiley & Sons (1987).

15. *Computerized Quantitative Infrared Analysis*, A Symposium Sponsored By ASTM Committee E-13 On Molecular Spectroscopy and Federation of Analytical Chemistry and Spectroscopy Societies (FACSS), Philadelphia, Pa, Philadelphia, Pa, ASTM (1987).

16. Osborne B. G., *Near Infrared Spectroscopy in Food Analysis*, Harlow, Essex, England, Longman Scientific & Technical; New York, John Wiley & Sons (1986).

17. Nakamoto K., *Infrared and Raman Spectra of Inorganic and Coordination Compounds*, 4th ed., New York, John Wiley & Sons (1986).

18. Griffiths P. R., *Fourier Transform Infrared Spectrometry*, New York, John Wiley & Sons (1986).

19. *Chemical, Biological and Industrial Applications of Infrared Spectroscopy*, New York, John Wiley & Sons (1985).

20. *Infrared Methods for Gaseous Measurements: Theory and Practice*, New York, John Wiley & Sons (1985).

21. *Oils, Lubricants, and Petroleum Products: Characterization by Infrared Spectra*, New York, John Wiley & Sons (1985).

22. Reidel D., *Fourier Transform Infrared Spectroscopy: Industrial Chemical and Biochemical Applications*, Sold and Distributed in the U.S.A. and Canada by Kluwer Academic Publishers (1984).

23. Martin A. E., *Infrared Interferometric Spectrometers*, Amsterdam, New York, Elsevier Scientific Pub. Co., New York: Distributors for the U.S. and Canada, Elsevier North Holland (1980).

24. Siesler H. W., *Infrared and Raman Spectroscopy of Polymers*, New York: M. Dekker (1980).

25. Socrates G., *Infrared Characteristic Group Frequencies*, Chichester; New York, Wiley (1980).

26. Holly S., *Absorption Spectra in the Infrared Region: Theoretical and Technical Introduction* / Budapest, Akademiai Kiado, 1975. 183 P., Ill.; 30 cm.

27. Smith A. L., *Applied Infrared Spectroscopy, Fundamentals, Techniques and Analytical Problem-Solving*, New York, John Wiley & Sons (1979).

28. Avram M., *Infrared Spectroscopy, Applications in Organic Chemistry*, Huntington, N.Y., R. E. Krieger Pub. Co. (1978).

29. De Faubert Maunder M. J., *Practical Hints on Infrared Spectrometry From a Forensic Analyst*, London, (31 Camden Rd., Nw1 9LP), Adam Hilger Ltd.

30. Avram M., *Infrared Spectroscopy: Applications in Organic Chemistry*, New York, Wiley-Interscience (1972).

31. *Fourier Transform Infrared Spectroscopy, Applications to Chemical Systems* / New York, Academic Press (1984).

32. Nakamoto K., *Infrared and Raman Spectra of Inorganic and Coordination Compounds*, 3d ed., New York, John Wiley & Sons (1978).

33. *Infrared, Correlation, and Fourier Transform Spectroscopy*, New York, M. Dekker (1977).

34. Scheinmann F., *An Introduction to Spectroscopic Methods for the Identification of Organic Compounds*, 1st ed., Oxford, New York, Pergamon Press (1970–74).

35. Dolphin D., *Tabulation of Infrared Spectral Data*, New York, John Wiley & Sons (1977).

Compilations of Infrared Spectra

1. *The Handbook of Infrared and Raman Characteristic Frequencies of Organic Molecules*, Boston, Academic Press (1991).

2. *The Infrared Spectra Atlas of Surface Active Agents*, Philadelphia, Pa., Sadtler Research Laboratories, a division of Bio-Rad Corporation (1989).

3. Socrates G., *Infrared Characteristic Group Frequencies: Tables and Charts*, 2nd ed., Chichester; New York, John Wiley & Sons (1994).

4. Jones G. C., *Infrared Transmission Spectra of Carbonate Minerals*, 1st ed. London; New York, Chapman & Hall (1993).

5. Buback, Michael, *FT-NIR Atlas*, New York, VCH (1993).

6. Garton A., *Infrared Spectroscopy of Polymer Blends, Composites and Surfaces*, Munich; New York, Hanser Publishers; New York, Distributed in The U.S.A. and Canada By Oxford University Press (1992).

7. *The Infrared Spectra Atlas of Coating Chemicals*, Philadelphia, Pa., Sadtler Research Laboratories, a division of Bio-Rad Corporation (1989).

8. Keller R. J., *The Sigma Library of FTIR Spectra*, 1st ed., Sigma Chemical Co. (1986).

9. Pouchert C. J., *The Aldrich Library of FT-IR Spectra*, 1st ed., Milwaukee, Wis., Aldrich Chemical Co. (1986).

10. *Regulated and Major Industrial Chemicals: A Special Collection of Infrared Spectra*, Kirkwood, Mo, U.S.A., The Society (1983).

11. *The Coblentz Society Desk Book of Infrared Spectra*, 2nd ed., Kirkwood, Mo., The Society (1982).

12. *The Sadtler Infrared Spectra Handbook of Minerals and Clays*, Philadelphia, Pa., Sadtler Research Laboratories, a division of Bio-Rad Corporation (1982).

13. *Vibrational Intensities in Infrared and Raman Spectroscopy*, Amsterdam, New York, Elsevier Scientific Pub. Co. (1982).

14. *The Infrared Spectra Handbook of Priority Pollutants and Toxic Chemicals*, Philadelphia, Pa., Sadtler Research Laboratories, a division of Bio-Rad Corporation (1982).

15. *The Sadtler Infrared Spectra Handbook of Esters*, Philadelphia, Pa., Sadtler Research Laboratories, a division of Bio-Rad Corporation (1982).

16. Coblentz Society, *Gases and Vapors:* A special collection of evaluated infrared spectra from the Coblentz Society, inc., with partial support from the Joint Committee on Atomic & Molecular Physical Data and the Office of Standard Reference Data, Kirkwood, Mo: The Society (1980).

17. Coblentz Society, *Plasticizers and Other Additives:* A special collection of infrared spectra from The Coblentz Society, Inc., 2d ed., Kirkwood, Mo., The Society (1980).

18. Coblentz Society, *Halogenated Hydrocarbons*, A special collection of infrared spectra from the Coblentz Society, Inc., 3rd ed. Kirkwood, Mo., The Society (1984).

19. Coblentz Society, *The Coblentz Society Desk Book of Infrared Spectra*, Kirkwood, Mo., The Society (1977).

20. Van Der Marel H. W., *Atlas of Infrared Spectroscopy of Clay Minerals and Their Admixtures*, Amsterdam; New York, Elsevier Scientific Pub. Co. (1976).

21. *The Infrared Spectra of Minerals*, London, Mineralogical Society (1974).

22. Gadsden J. A., *Infrared Spectra of Minerals and Related Inorganic Compounds*, London, Butterworths (1975).

Computer Tutorial Programs

IR Simulator, Paul F. Schatz, University of Wisconsin—Madison, distributed by Trinity Software, Compton, New Hampshire. Both the MS DOS and MACINTOSH versions are available. These programs emulate the Perkin–Elmer 1310 Infrared Spectrophotometer.

Spectradeck®, Volume 1. Paul F. Schatz, University of Wisconsin—Madison, distributed by Trinity Software, Compton, New Hampshire. MACINTOSH version only. Contain 40 compounds in a Hypercard Stack. The program aids in the understanding and analysis of IR, MS, NMR and ^{13}CMR spectra.

IR Tutor, Charles B. Abrams in collaboration with the Perkin-Elmer Corporation, Windows version only. Distributed by the Perkin-Elmer Corporation. This is a highly recommended program for understanding the origin of characteristic group frequencies (1992–93).

Appendix

A.1 Infrared Group Frequencies (cm^{-1})

Abbreviations Used

s = strong   m = medium   w = weak   v = variable

A.1.1 Infrared Solvents

Solvent	Indeterminate regions
Hexachloro-1,3-butadiene	1610–1562, 1191 to 1150, 1000 to 910 and 870 to 770
Carbon disulfide	1666 to 1333, 2325 to 2040, 885 to 826 and 794 to 746
Carbon tetrachloride	1588 to 1493, 854 to 694, 1250 to 1191 and 1020 to 960
Chloroform	2940, 2352, 1282 to 1136, 1539 to 1370, 1075 to 1010 and 927 to 910, 833 to 724

A.1.2 Bands Related to Sample Preparation

Method of determination	Indeterminate regions
Liquid films	None
Salt pellets	None
Nujol mull (mineral oil)	Strong absorptions centered at 2987, 1466 and 1377

A.2 General Infrared Assignments

Region	Assignment
3772 to 2670	O–H, N–H and C–H stretching
800 to 600	C–Cl stretching (s)
1430 to 1000	C–F stretching (very sharp)
1640 to 1150	N–H, C–H and O–H bond
1350 to 800	C–O, C–N and C–C stretching
900 to 588	C–H and N–H rocking

A.3 Carbonyl Absorptions

A.3.1 Ketones (All Strong Absorptions)

Saturated, acyclic	1715
Saturated, cyclic	1715
Six-membered ring (and higher)	1712
Five-membered ring	1745
Four-membered ring	1745
Three-membered ring	near 1880
Aromatic, mono	1690
Aromatic, di	1665
α, β -unsaturated, acyclic	1675
α, β -α', β'-unsaturated, acyclic	1666

A.3.2 α, β-Unsaturated, Cyclic Ketones (cm^{-1})

Six-membered ring	1667
Five-membered ring	1705
α-Diketones, saturated, acyclic	720
β-Diketones (enolic), saturated, acyclic	1588
Diketones, saturated, acyclic	1715
Quinones, two carbonyls in one ring	1672
Quinones, two carbonyls in two rings	1645

A.3.3 Aldehydes (All Strong Absorptions)

Saturated, aliphatic	1730
α, β-Unsaturated, aliphatic	1690
α, β, γ, δ-Unsaturated, aliphatic	1670
Aryl	1706
C–H stretching vibrations	2900 to 2695 (w), usually two bands with one band near 2710
C–H deformation vibrations	970 to 788 (medium)

A.3.4 Carboxylic Acids (All Strong Absorptions) (cm^{-1})

Saturated aliphatic	1715
α, β-Unsaturated aliphatic	1700
Aryl	1697
Hydroxyl stretching vibrations	2700 to 2500 (W)
Hydroxyl deformation vibrations	952 to 903 (W)
Carboxylate anion	1575 and 1360 (S)

A.3.5 Amides

A.3.5.1 Carbonyl Vibrations (Amide I Band, All Strong) (cm^{-1})

Primary	1650
Secondary	1684
Tertiary	1652
Cyclic, α-lactams and larger	1678
Cyclic, α-lactams, fused to another ring	1725
Cyclic, α-lactams	1700
Cyclic, β-lactams and larger	1745
Cyclic, β-lactams, fused to another ring	1774

A.3.5.2 N–H Stretching Vibration (All Medium)

Primary, free	3500 to 3400
Primary, bonded	3345 to 3185
Secondary, free	3425
Secondary, bonded	3228

A.3.6 N–H Deformation (Amide II Band, All Strong)

Primary	1634
Secondary	1570

A.3.7 Esters Carbonyl (All Strong)

Saturated acyclic	1742
Saturated cyclic	
δ-Lactones (and larger rings)	1742
α-Lactones	1770
β-Lactones	1820
Unsaturated	
α, β-Unsaturated acyclic and aryl	1722
α, β-Unsaturated-δ-lactones	1722
α, β-Unsaturated-γ-lactones	1752
β, γ-Unsaturated-γ-lactones	1800
α-Ketoesters	1748
β-Ketoesters (enolic)	1650
γ-Ketoesters (and higher)	1740

A.3.8 Anhydride Carbonyl (All Strong)

Acyclic; unconjugated, also cyclic	**(cm^{-1})**
Six-membered rings and higher	1825 and 1766
Acyclic, conjugated	1720 and 1715
Cyclic, five-membered (unconjugated)	1845 and 1773
Cyclic, five-membered (conjugated)	1790 and 1692

A.3.9 Hydrocarbon Chromophore

C–H stretching vibrations	
Alkane	2919 (M–S)
Ethylene, monosubstituted (vinyl)	3020 and 3088
Ethylene, disubstituted (trans)	3020 and 3088
Ethylene, disubstituted (cis)	3020 and 3088
Ethylene, trisubstituted	3088
Acetylene	3321 (M)
Aromatic	3030 (M)

C–H deformation vibrations

Alkane, C–H	1339 (W)
Alkane, C–CH$_3$	1450 (M), 1337 (S)
Alkane, –CH$_2$–	1456 (M)
Alkane, isopropyl	1380 (S), 1368 (S)
Alkane, tert-butyl	1389 (M), 1365 (S)
Ethylene, monosubstituted (vinyl)	990 (S), 910 (S) and 1414 (S)
Ethylene, disubstituted (trans)	3020 and 3088
Ethylene, disubstituted (cis)	3020 and 3088
Acetylene	3020 and 3088

A.3.10 Substitution Patterns on Aromatic Ring (cm^{-1})

5 adjacent free H atoms	770–730 (S), 710–690 (S), 1055 (W), 1085 (W) and 1036 (W)
4 adjacent free H atoms	770–735 (S), 1195 (V) 1105 (W), and 1036 (W)
3 adjacent free H atoms	810–750 (s), 1195 (v), 1105 (w) and 1036 (w)
2 adjacent free H atoms	860–800 (s), 1055 (w), 1036 (w) and 1086 (w)
1 adjacent free H atoms	900–860 (m)

A.3.11 C–C Multiple Bond Stretching (cm^{-1})

Ethylenic, non conjugated	1645 (v)
Ethylenic, aryl conjugated	Near 1623 (s)
Ethylenic, carbonyl or ethylenic conjugated near	1600 (s)

For simple olefins, the degree of substitution has approximately the following effect (all medium absorptions)

Monosubstituted (vinyl)	Near 1645
Disubstituted, trans	Near 1675
Disubstituted, cis	Near 1659
Disubstituted, ethylidene	Near 1652
Trisubstituted	Near 1670
Tetrasubstituted	Near 1670 (w)

A.3.12 Acetylenic Types

Acetylenic, monosubstituted	Near 2120 (s)
Acetylenic, disubstituted	2192 (v)
Allenic	Near 1950 and 1058 (both m)

A.3.13 Aromatic Skeletal-in-Plane Vibrations

Skeletal in-plane vibrations are found near 1600 (v) and 1500 (v) and for conjugated rings, near 1580 (m). Aromatic fine structure bands are found as a series of bands of weak intensity between 2000 and 1666.

A.4 Miscellaneous Compounds (Proximate Bands)

A.4.1 Alcohol and Phenols (Stretching) (cm^{-1})

Free O–H	3610 (v, sharp)

Intermolecular hydrogen bonded (broken on dilution)

Single bridge compounds	3495 (v, sharp)
Polymeric association	3300 (s, broad)

Intramolecular hydrogen bonded (no change on dilution)

Single bridge compounds	3450 (v, sharp)
Chelate compounds	3220 to 2500 (w, very broad)

Deformations

Primary alcohols	Near 1053 (s) and 1350 to 1250 (s)
Secondary alcohols	Near 1053 (s) and 1350 to 1250 (s)
Tertiary alcohols	Near 1150 (s) and 1410 to 1300 (s)
Phenols	Near 1205 (s) and 1410 to 1300 (s)

A.4.2 Amines, N–H Stretching

Primary amines, two bands	3495 (m)
Secondary amines, one band	3390 (m)
Imines, one band	3290 (m)
Amine salts, –NH$_3$	3080 (m)

A.4.3 Unsaturated Nitrogen Compounds

A.4.3.1 Nitriles, −C≡N stretching

Saturated alkyl nitriles	2252 (S)
Aryl nitriles	2228 (s)
α, β-unsaturated alkyl nitriles	2232 (S)

A.4.3.2 C=N− Stretching (cm^{-1})

Open chain compounds	1665 (v)
Open chain α, β-unsaturated compounds	1655 (v)
Conjugated cyclic compounds	1588 (V)
N = N− stretching	1600 (v)
N = N stretching (azides)	1250 (w) and 2140 (s)

A.4.3.3 Nitro Compounds

Aromatic	1520 (m) and 1325 (m)
Aliphatic	Usually at slightly lower wavelengths that the aromatic counterparts

A.4.3.4 Nitroso Compounds

O−N=O (nitrites)	1666 (s) and 1628 (s)
C−N=O	Near 1562 (s)
N−N=O	Near 1450 (s)

A.4.3.5 Halogen Compounds C−X Stretching

C−F	1430 to 1000 (very sharp)
C−Cl	800 to 600 (s)
C−Br	600 to 500 (s)
C−I	Near 500 (s)

A.4.3.6 Sulfur Compounds

−S−H	1540 (w)
−S−S−	500 to 400
C−S	650 (w)
N−C=S	1485 (s)

S=O (sulfoxides)**	1055 (s)
SO_2 (sulfones)	1150 (s) and 1325 (s)
SO_2H (sulfinic acids)	Near 1090 (s)
SO_3H (sulfonic acids)	1170 (s), 1045 (s) and near 650 (s)

A.5 Characteristic Overtone Patterns for Substitutions on Benzene

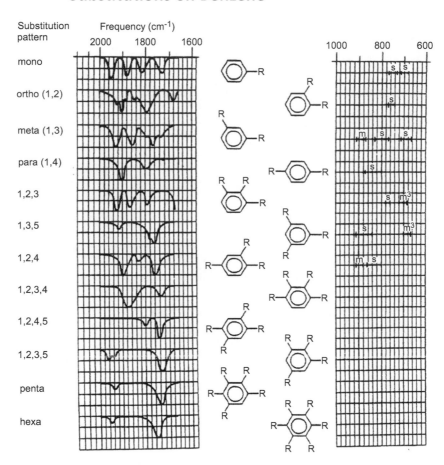

Note: Absorption patterns for the substituted benzenes (s = a strong peak, m = a medium peak, m^3 = three medium peaks).

A.6 List of Infrared Symbols and Abbreviations

Mode of Vibration	Symbols used	Others (Lit.)
Stretch	st	υ
Asymmetric stretch	st_{as}	υ_{as}
Symmetric stretch	st_s	υ_a
In-plane stretch		υ_β
Out-of-plane stretch		υ_γ
Deformation (bend, twist, rock, wag, etc.)	d	δ, δ' (bend, twist, or rock)
Asymmetric deformation (usually bend)	d_{as}	
Symmetric deformation (usually bend)	d_s	
In-plane bending	b	δ, R
Out-of-plane bending	b	γ
Asymmetric bending	b_{as}	δ_{as}
Symmetric bending	b_s	δ_s
Scissor	s	
Wag (out-of-plane)	w	γ
Rock (in-plane)	r	ρ, r_β
Rock (out-of-plane)	r	r_γ
Twist (out-of-plane)	t	t
Torsion	t	τ
Intensity		
Weak		w
Medium		m
Strong		s
Very strong		vs

A.7 The Infrared Spectra of Some Common Chemicals Found in the Organic Chemistry Laboratory

Styrene 1	Ethyl alcohol 2	Benzoyl chloride 3	Nitromethane 4
Polystyrene 5	Phenol 6	Neopentyl alcohol 7	Urea 8
Adipic acid 9	Acetyl chloride 10	Butyl amine 11	Acetic anhydride 12
Acetone 13	2-Hexanone 14	Cumene 15	Carbon tetrachloride 16
Dichloromethane 17	Chloroform 18	Isopropyl alcohol 19	Mineral oil 20
Aniline 21	Acetophenone 22	Benzaldehyde 23	Diethyl ether 24
Nitrobenzene 25	Silicon oil 26	1-Heptene 27	Allyl benzene 28
Ethylene glycol 29	phenyl ether 30	Furan 31	Anisole 32
p-Dichlorobenzene 33	Vinyl bromide 34	1,3-Cyclohexadiene 35	Naphthalene 36

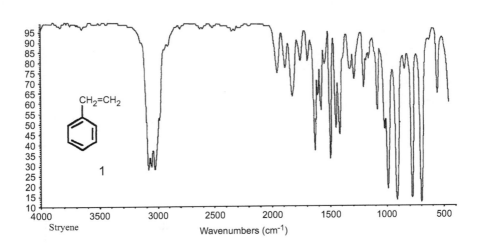

Stryene

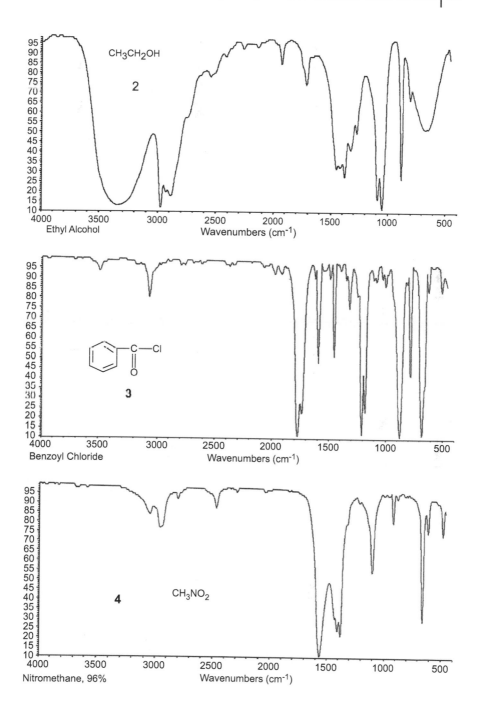

CH₃CH₂OH

2

Ethyl Alcohol

Wavenumbers (cm⁻¹)

3

Benzoyl Chloride

Wavenumbers (cm⁻¹)

4   CH₃NO₂

Nitromethane, 96%

Wavenumbers (cm⁻¹)

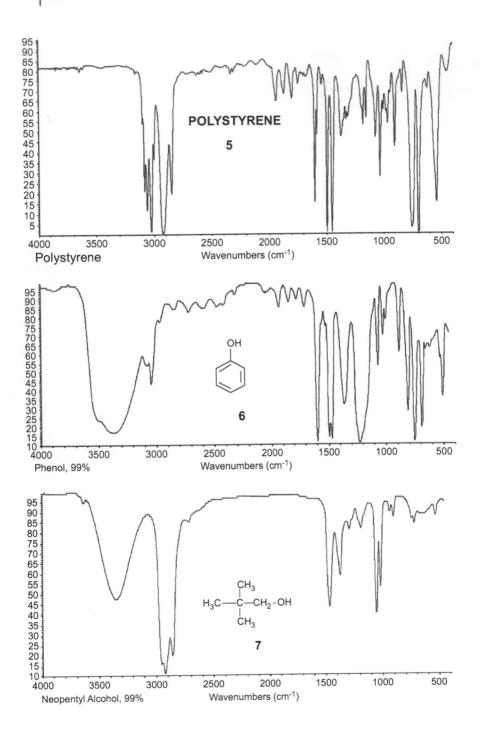

Polystyrene

POLYSTYRENE
5

Wavenumbers (cm⁻¹)

Phenol, 99%

6

Wavenumbers (cm⁻¹)

Neopentyl Alcohol, 99%

7

Wavenumbers (cm⁻¹)

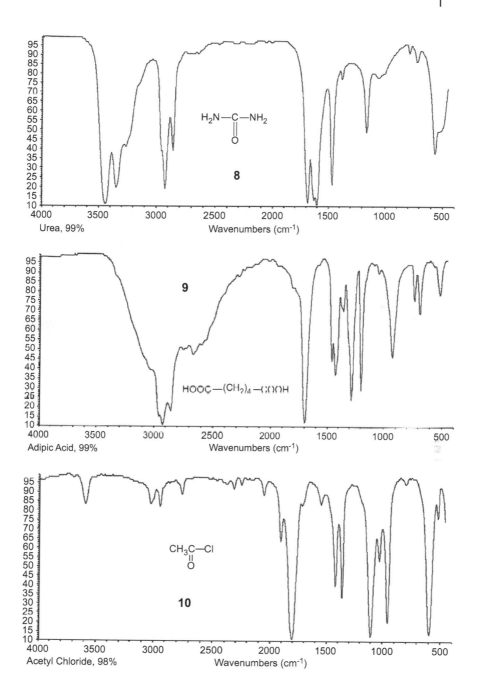

Urea, 99%   Wavenumbers (cm⁻¹)

8

Adipic Acid, 99%   Wavenumbers (cm⁻¹)

9

HOOC—(CH₂)₄—COOH

Acetyl Chloride, 98%   Wavenumbers (cm⁻¹)

CH₃C—Cl
‖
O

10

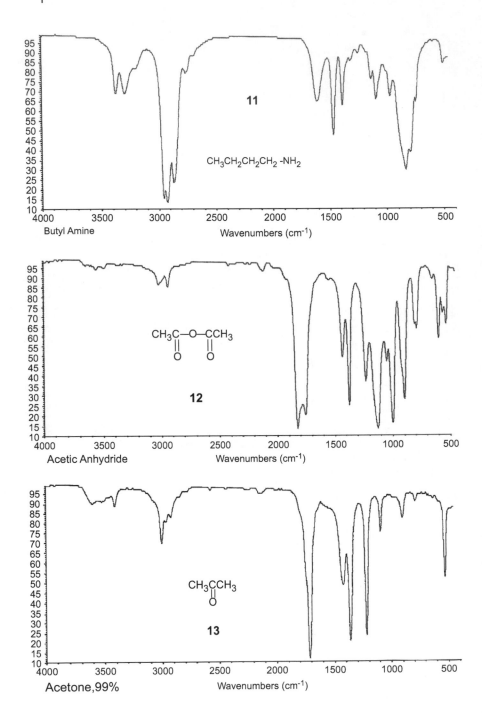

11

CH₃CH₂CH₂CH₂ -NH₂

Butyl Amine

Wavenumbers (cm⁻¹)

CH₃C—O—CCH₃

12

Acetic Anhydride

Wavenumbers (cm⁻¹)

CH₃CCH₃

13

Acetone,99%

Wavenumbers (cm⁻¹)

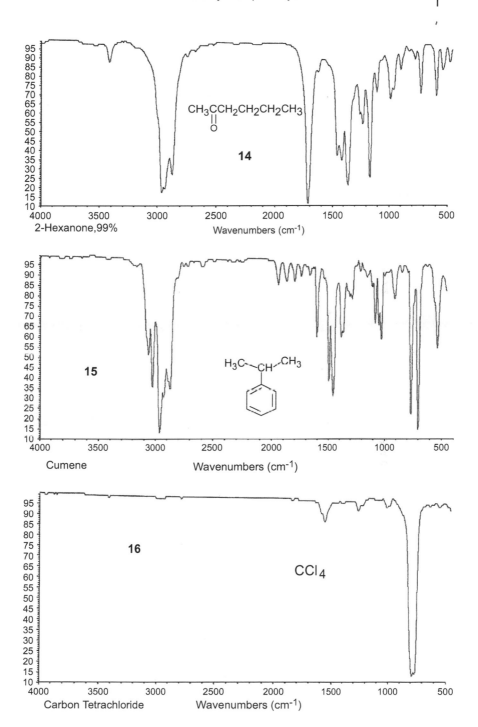

$$CH_3CCH_2CH_2CH_2CH_3$$
$$O$$

14

2-Hexanone,99%

Wavenumbers (cm⁻¹)

15

$$H_3C \quad CH_3$$
$$CH$$

Cumene

Wavenumbers (cm⁻¹)

16

$$CCl_4$$

Carbon Tetrachloride

Wavenumbers (cm⁻¹)

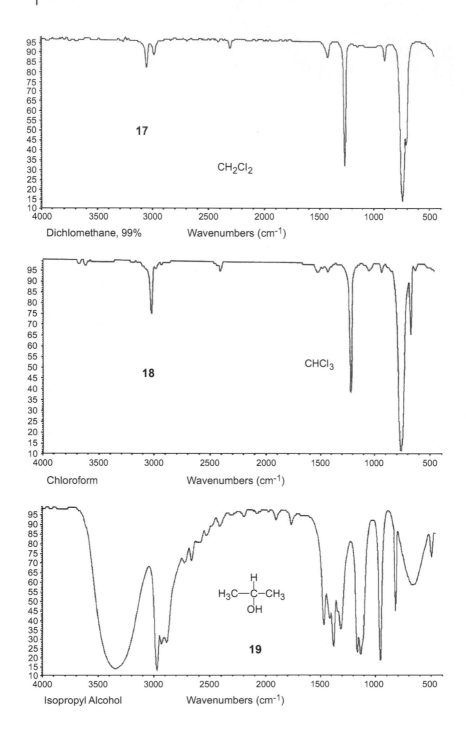

17

CH$_2$Cl$_2$

Dichlomethane, 99%   Wavenumbers (cm^{-1})

18

CHCl$_3$

Chloroform   Wavenumbers (cm^{-1})

H$_3$C—C—CH$_3$
H
OH

19

Isopropyl Alcohol   Wavenumbers (cm^{-1})

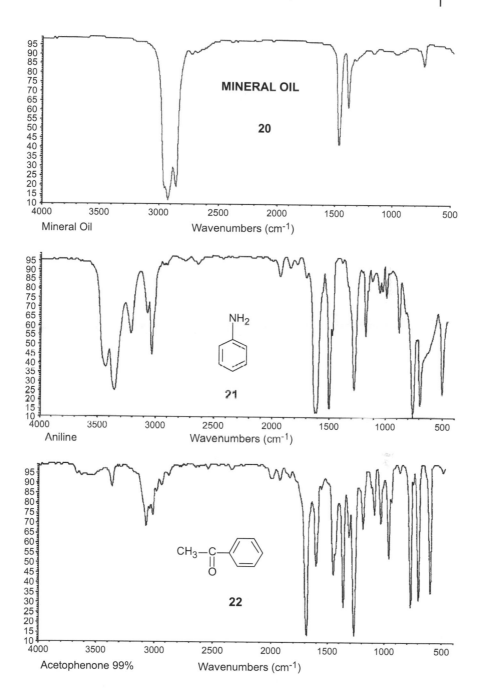

Mineral Oil

MINERAL OIL

20

Wavenumbers (cm⁻¹)

NH₂

21

Aniline

Wavenumbers (cm⁻¹)

CH₃—C

22

Acetophenone 99%

Wavenumbers (cm⁻¹)

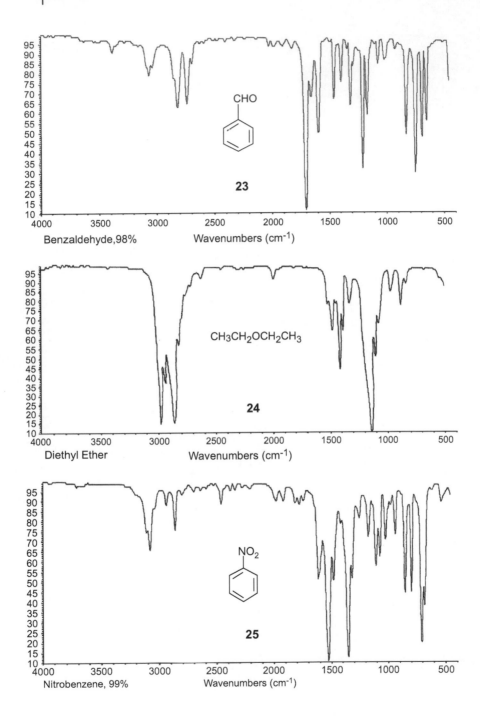

23

CHO

Benzaldehyde,98% Wavenumbers (cm⁻¹)

24

CH₃CH₂OCH₂CH₃

Diethyl Ether Wavenumbers (cm⁻¹)

25

NO₂

Nitrobenzene, 99% Wavenumbers (cm⁻¹)

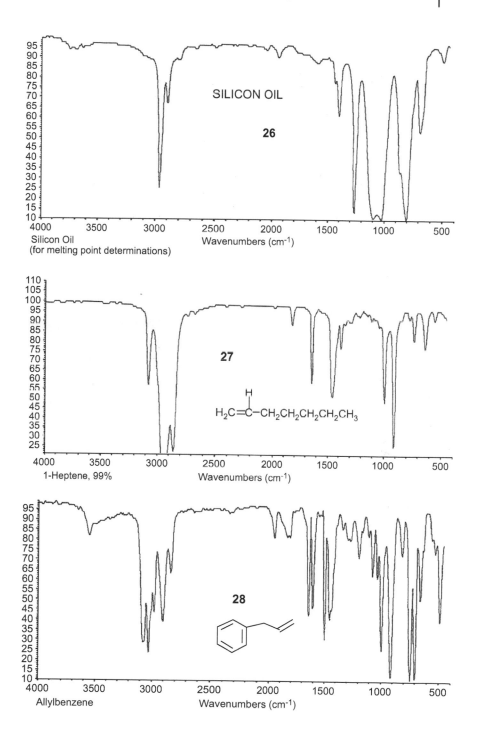

SILICON OIL

26

Silicon Oil
(for melting point determinations)
Wavenumbers (cm⁻¹)

27

H₂C=C—CH₂CH₂CH₂CH₂CH₃
|
H

1-Heptene, 99%
Wavenumbers (cm⁻¹)

28

Allylbenzene
Wavenumbers (cm⁻¹)

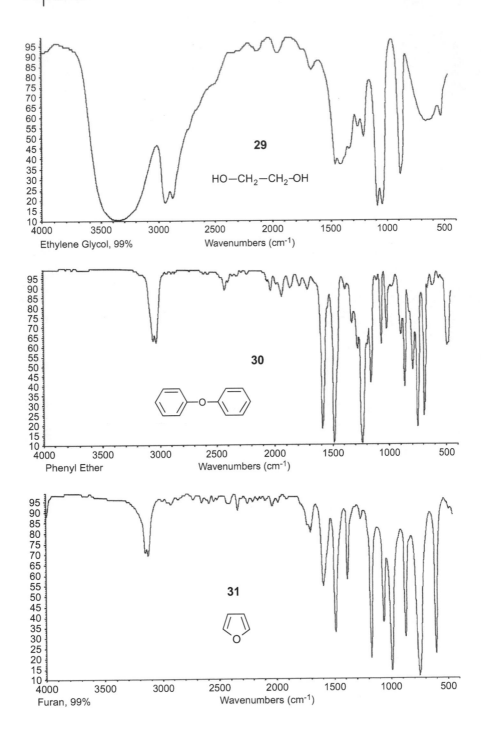

29

HO—CH$_2$—CH$_2$-OH

Ethylene Glycol, 99%

Wavenumbers (cm^{-1})

30

Phenyl Ether

Wavenumbers (cm^{-1})

31

Furan, 99%

Wavenumbers (cm^{-1})

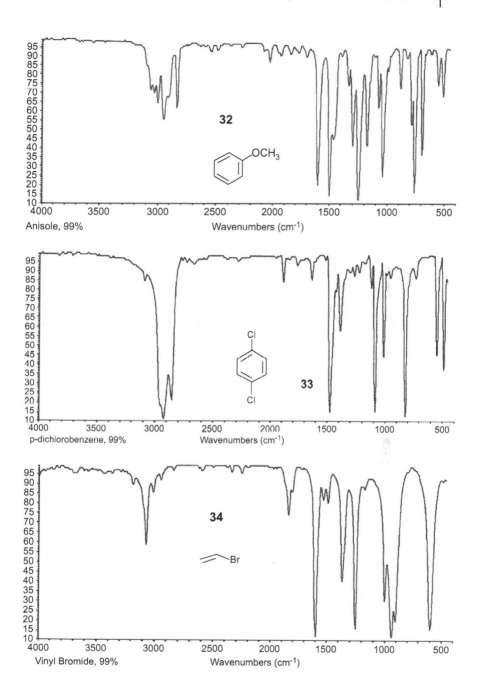

Anisole, 99%   Wavenumbers (cm⁻¹)

32

p-dichlorobenzene, 99%   Wavenumbers (cm⁻¹)

33

Vinyl Bromide, 99%   Wavenumbers (cm⁻¹)

34

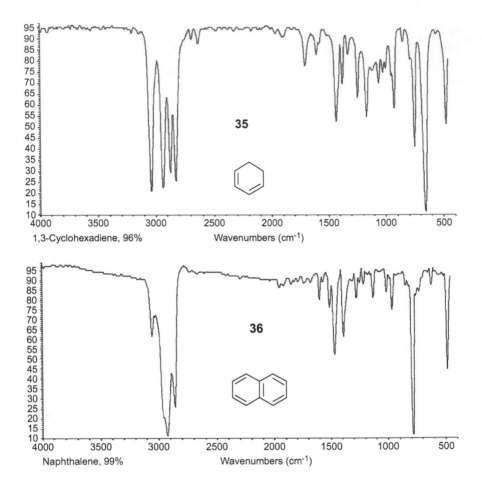

1,3-Cyclohexadiene, 96%   Wavenumbers (cm⁻¹)

Naphthalene, 99%   Wavenumbers (cm⁻¹)

Index

Printed and bound by CPI Group (UK) Ltd, Croydon, CR0 4YY

23/10/2024

0177667-0002